Elias Loomis

A Treatise on Meteorology

With a Collection of Meteorological Tables

Elias Loomis

A Treatise on Meteorology
With a Collection of Meteorological Tables

ISBN/EAN: 9783337217655

Printed in Europe, USA, Canada, Australia, Japan

Cover: Foto ©berggeist007 / pixelio.de

More available books at **www.hansebooks.com**

A

TREATISE

ON

METEOROLOGY.

WITH A COLLECTION OF

METEOROLOGICAL TABLES.

BY ELIAS LOOMIS, LL.D.,

PROFESSOR OF NATURAL PHILOSOPHY AND ASTRONOMY IN YALE COLLEGE, AND AUTHOR OF A
"COURSE OF MATHEMATICS."

NEW YORK:
HARPER & BROTHERS, PUBLISHERS,
327 TO 335 PEARL STREET,
FRANKLIN SQUARE.
1868.

PREFACE.

WITHIN the past forty years a vast amount of meteorological observations has been accumulated from almost every part of the world, and particularly from the United States. Within the limits of our own country we have observations, more or less extensive, from more than a thousand stations, and some of these registers are very accurate and complete. So great an amount of labor expended upon observations ought certainly to lead to some valuable results. Such results have already been in part attained, but they are generally published in very large works, or in elaborate memoirs whose object is limited to the discussion of special questions. Many of these memoirs are only to be found in foreign languages, and nearly all of them are too elaborate to circulate freely even among the mass of tolerably intelligent observers. It will probably be conceded that there has not hitherto appeared, at least in the English language, any general treatise on Meteorology which furnishes a comprehensive view of the present condition of every branch of this science with a minuteness sufficient to satisfy one who is himself engaged in the business of observing. In the present volume an attempt has been made to furnish a concise exposition of the principles of Meteorology in a form adapted to use as a text-book for instruction, and at the same time to exhibit the most important results of recent researches. That this attempt has been but partially successful no one can be more fully aware than the author; nevertheless it is hoped that this volume will compare favorably with any work which has hitherto appeared having the same objects in view. This treatise has been in contemplation for many years, during which I have been collecting materials for this purpose.

It would have been quite easy to have expanded the book to double its present size, and in such a form it might have been more satisfactory to those who are themselves engaged in original researches; but I have aimed to prepare a work which should not only be useful to observers, but should also be adapted to purposes of instruction in our colleges and scientific schools. It is hoped that this volume may serve to stimulate observers, by showing them the important results already deduced from their labors, and also by calling their attention to the unsettled problems which require for their solution either more accurate or more numerous observations.

I have again to acknowledge my obligations to Professor H. A. Newton, who has read all the proofs of this work, and to whom I am indebted for numerous suggestions, particularly in the last chapter, which relates to a subject to which he has devoted special attention.

CONTENTS.

CHAPTER V.

PRECIPITATION OF THE VAPOR OF THE AIR.

SECTION I.—DEW.

SECTION II.—HOAR-FROST.

SECTION III.—FOG.

SECTION IV.—CLOUDS.

SECTION V.—RAIN.

SECTION VI.—SNOW.

SECTION VII.—HAIL.

CHAPTER VI.

STORMS, TORNADOES, AND WATER-SPOUTS.

SECTION I.—THEORY AND LAWS OF STORMS.

SECTION II.—CYCLONES.

SECTION III.—TORNADOES.

SECTION IV.—PILLARS OF SAND AND WATER-SPOUTS.

SECTION V.—PREDICTIONS OF THE WEATHER.

TABLES.

METEOROLOGY.

CHAPTER I.

CONSTITUTION AND WEIGHT OF THE ATMOSPHERE.

1. THE term *meteor* was formerly employed to denote those natural phenomena which occur within the limits of our atmosphere, as the wind, rain, thunder, the rainbow, etc.; and *Meteorology* might, therefore, be defined as that branch of Natural Philosophy which treats of Meteors.

This branch of science treats of the constitution and weight of the air; of its temperature and moisture; of the movements of the atmosphere; of the precipitation of vapor in the form of dew, hoar-frost, fog, cloud, rain, snow, and hail; of the laws of storms, including tornadoes and water-spouts; with various electrical phenomena, including atmospheric electricity, thunderstorms, and the Polar Aurora; as also various optical phenomena, including the rainbow, twilight, mirage, coronæ, and halos; to which are generally added aerolites and shooting stars.

2. *Composition of the Air.*—Atmospheric air is not a simple substance, as was once believed, but consists of nitrogen and oxygen, together with more or less vapor of water, and almost always a little carbonic acid. The nitrogen and oxygen are combined in the ratio of 79.1 to 20.9 by volume. These proportions are generally the same in all parts of the globe, and at all accessible elevations above the earth's surface. During a balloon ascent, air has been collected from an elevation of 21,774 feet, and its constitution was found to be sensibly the same as that of air at the earth's surface.

Atmospheric air contains a little carbonic acid (from 0.0004 to 0.0006 in the open country), and a variable amount of vapor of water. The amount of moisture in the atmosphere sometimes forms four per cent. of its entire weight, and sometimes is less than a tenth of one per cent.

3. *Distinction between Vapors and Gases.*—Aeriform bodies are naturally divided into two classes. Some are easily reduced to the liquid state, and are called *vapors*, as the vapor of water. Others always remain in the aeriform state, or can only be reduced to the liquid state with the greatest difficulty. These are called *gases*, such as oxygen, nitrogen, hydrogen, etc.

4. *Law of Mixture of Gases.*—When vapors and gases are superposed upon each other, they obey a law different from liquids. If we pour into the same vessel several liquids which exert no chemical action upon each other, they will arrange themselves in the order of their specific gravities; the heaviest will subside to the bottom, and the lightest will float upon the surface. But if we introduce into the same vessel different gases, they will not arrange themselves in separate strata in the order of their specific gravities, but will mutually penetrate each other, and after a short time the proportion of the several gases will be the same in every part of the vessel. This movement of gases toward each other has received the name of *diffusion*.

5. *Dalton's Theory of the Atmosphere.*—According to the theory of Dalton, the gases which compose the atmosphere are not in a state of chemical combination, but the particles of either gas have neither attraction nor repulsion for those of another, and each of them is disposed precisely as if the others were not present. He therefore considered that the earth is surrounded in effect by four atmospheres, which interpenetrate each other, but without interference.

The hypothesis that there is no repulsion between the particles of the different gases which compose the atmosphere has not been generally received. The diffusion of gases may be explained by supposing that the molecules of gases are situated at great distances from each other; and each gas, therefore, presents vast pores through which the particles of the other gas may penetrate.

6. *Gases in the upper Regions of the Atmosphere.*—In the upper and inaccessible regions of the atmosphere there are no other gases than those found at the surface of the earth, for such gases would in time penetrate to the earth's surface by the force of diffusion. The hypothesis, therefore, which explains certain fiery meteors by the assumption of an inflammable gas in the upper regions of the atmosphere, is inadmissible.   ✦

7. *Proportions of the Gases at Great Elevations.*—A stratum of air near the earth sustains the weight of the entire superincumbent atmosphere, and its density is thereby increased. This density diminishes as we rise above the surface of the earth; and since each gas is distributed as if no other gas was present, this diminution (which depends upon the weight of the gas) ought not to be the same for each of the constituents of the atmosphere. At great elevations, the proportion of these gases should therefore be different from what it is at the earth's surface. It has been computed that, at the height of four miles, the proportion of nitrogen to oxygen should be one per cent. greater than at the earth's surface. Observation has, however, shown that there is no such difference, a result which is attributed to the constant agitation of the atmosphere, by which the different strata are thoroughly mingled together.

8. *Limit of the Atmosphere determined by Centrifugal Force.*—Since the earth's attraction, which retains the air near to its surface, varies inversely as the square of the distance from the centre, while the centrifugal force arising from the earth's rotation increases with the distance, there must be a certain height at which these two forces are equal, and beyond this distance the air will be dissipated by centrifugal force. This point is about 25,000 miles from the earth's centre.

9. *Estimate of the Actual Height of the Atmosphere.*— Other considerations indicate a much lower limit to the atmosphere. The atmosphere must terminate at that height where the attraction of the earth is just equal to the repulsion between the particles of air, and this repulsion is diminished by the low temperature of the upper regions.   At the height of 50 miles the atmosphere is well-nigh inappreciable in its effect upon twilight. The phenomena of lunar eclipses indicate that the earth's atmosphere is appreciable at the height of 66 miles. The phenomena of shooting

stars and the auroral light indicate that an appreciable atmosphere exists at the height of 200 or 300 miles, and probably
more than 500 miles from the earth's surface.

10. *Construction of the Barometer.*—The weight of the atmosphere
is measured by a barometer. If we take a glass
tube, A B, about three feet in length, hermetically
sealed at one end and open at the other, fill
it with quicksilver, and then, closing the open
end of the tube with the finger, invert the tube,
and immerse the lower end in a cup filled with
mercury, on removing the finger the liquid will
fall only a moderate distance, and will be maintained at an elevation of about thirty inches above
the level of the liquid in the cup. The column
of mercury in the tube C D is supported by the
pressure of the air acting on the surface of the
mercury in the cup; and we conclude that the
weight of a column of mercury having a height
of thirty inches is equal to that of a column of
air of the same base, extending to the top of the atmosphere.
Such an instrument is called a *Barometer.* The barometer measures, therefore, the pressure of the air, and, in order to ascertain
its amount, we have only to attach to the glass tube a graduated
scale.

In order to allow entire freedom of motion to the column of
mercury, the diameter of the tube should not be too small. For
a stationary barometer, a tube having an internal diameter of half
an inch is not too great.

11. *How Air and Moisture are Excluded.*—Special care should be
taken to exclude from the tube both air and moisture, the presence of which would produce pressure upon the upper extremity
of the column of mercury, and depress it below its proper height.
It is found very difficult to attain this object perfectly. The tube
should be entirely clean, and the mercury should be filtered, and
both should be heated, in order to expel moisture. A small quantity of mercury is then introduced into the tube, special care being taken to prevent the admission of air-bubbles. The tube is
then held over a charcoal fire and heated until the mercury boils,

the tube being held in an inclined position, so that any particles which may adhere to the sides of the tube may easily escape. More mercury is now added, and the operation of boiling repeated as before, and thus the tube is gradually filled.

If a barometer-tube has been well freed from air and moisture, when the tube is suddenly inclined the mercury will strike the top of the tube with a sharp metallic sound.

12. *How the Height of the Column is Measured.*—The height of the mercury in the barometer varies from day to day, and the graduation of the scale by which its height is measured should have a sufficient range to comprehend the extreme variations in the height of the column. With a stationary barometer, these variations are generally comprehended between 27 and 31 inches. This portion of the scale is divided into tenths of an inch, and these spaces are still farther subdivided by means of a vernier.

The graduated scale may be either *fixed* or *movable.* If the scale be fixed, a correction will be required for the oscillations of the mercury in the tube. Suppose, when the air is at its mean pressure, the lower extremity of the graduated scale just touches the surface of the mercury in the cistern. When the pressure diminishes, the mercury which descends from the tube fills the cistern to a greater height, and its level rises above the lower extremity of the scale. When the pressure increases, mercury from the cistern ascends into the tube, and its level is left below the extremity of the scale. Thus the lower extremity of the graduated scale alternately sinks below the level of the cistern, and rises above it, in neither of which cases is the true pressure of the atmosphere directly indicated. As, however, when we know the relative diameters of the tube and cistern, the variations of the level of the cistern may be easily computed, such a barometer may give accurate results; yet the inconvenience is entirely remedied by making the scale movable. In this case the lower extremity of the scale is made to terminate in an ivory point, which, by the motion of a screw, D, may at each observation be brought to exact coincidence with A, the surface of the mercury in the cistern.

Fig. 2.

In some barometers the scale is fixed, but the level of the mercury in the cistern may be adjusted to the extremity of the scale by means of a screw, B.

In order that observations made with different barometers may be comparable, corrections are required both for temperature and for capillary action.

13. *Correction for Temperature.*—Heat expands the column of mercury; that is, diminishes its specific gravity, and thus a greater height is required to produce a given pressure. Now, since the barometer is daily subjected to changes of temperature, variations in the height of the column do not necessarily indicate variations of pressure. Before we can decide whether there has been a change of pressure, we must compute the effect due to the change of temperature. For this purpose, we must know the temperature of the mercury at each observation; and, accordingly, a thermometer always accompanies a barometer, and is technically called the *attached thermometer.* At every observation of the barometer the attached thermometer should also be observed. For the purpose of comparison, all barometric observations should be reduced to a standard temperature, and the temperature generally agreed upon is that of melting ice. The expansion of mercury from the temperature of melting ice to that of boiling water is $\frac{1}{55}$ of its volume, which is about $\frac{1}{10,000}$th part for one degree of Fahrenheit's thermometer. In order, therefore, to reduce the observed height of the barometer to the height which would have been indicated if its temperature had been 32°, we must subtract the ten thousandth part of the observed altitude for each degree above the freezing point. If the temperature be below 32°, this correction must be added to the observed altitude. Tables have been computed, from which we may obtain, by mere inspection, the correction to be applied to the observed height of the barometer. See Table VIII., pages 258–259.

14. *Correction for Capillary Action.*—By capillary action the column of mercury in the tube is depressed below that height which would just balance the pressure of the air, and a correction must be added to the observed heights of the barometer in order to obtain the true pressure of the atmosphere. This correction varies with the diameter of the tube.

In a tube whose diameter is	0.10 inch .20 " .30 " .40 " .50 " .60 " .	the depression amounts to	0.140 inch. .058 " .029 " .015 " .008 " .004 "

15. *The Aneroid Barometer* is an instrument for measuring the pressure of the atmosphere by means of the elasticity of a plate of metal. It consists of a cylindrical brass box, about three inches in diameter and half an inch in height, the sides of which are made very thin, and which is hermetically sealed after the air has been partly exhausted from the interior. When the pressure of the atmosphere increases, the inclosed air is compressed, the capacity of the box is diminished, and the two flat ends approach each other. When the pressure diminishes, the ends resume their former position, in consequence of the expansion of the inclosed air. By means of a combination of levers, this motion of the ends of the box is communicated to a pointer, which travels over a graduated dial-plate, and the original motion is magnified, so that the index travels over a space of three inches, while the end of the box only moves the $\frac{1}{300}$th of an inch. This instrument has the advantage of extreme portability, and, when well made, will measure small deviations from the mean pressure within one or two hundredths of an inch. In observations requiring great accuracy, it should, however, be frequently compared with a standard mercurial barometer.

Fig. 3.

16. *Self-registering Barometers.*—In order to diminish the labor of frequent observations of the barometer, attempts have been made to render it self-registering. One of the best methods of accomplishing this object is by means of *photography*. The light of a lamp or gas-flame, A, is concentrated by means of a lens, B,

Fig. 4.

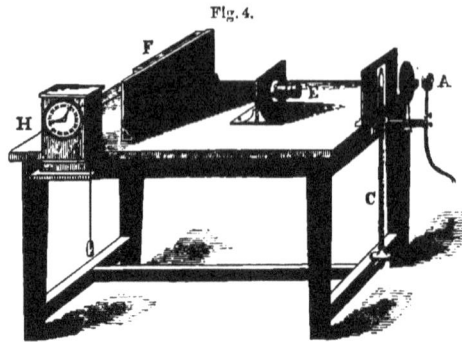

so as to strike upon the summit of the column of mercury in the barometer tube, C D. A sheet of paper suitably prepared for photographic experiments is attached to a frame, F, placed behind a screen, G, having a narrow vertical slit placed in the line of the rays passing through B. The mercury protects a portion of the paper from the action of the light of the lamp, while above the mercury the rays of the lamp fall unobstructed upon the paper. By means of a clock, H, the paper is carried steadily forward at the rate of about half an inch per hour, and thus the column of mercury leaves upon the paper a permanent record of its height for each instant of the day. At the close of the day a new

Fig. 5.

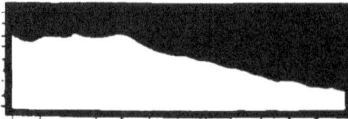

m't. 2h 4   6   8   10 noon 2h 4   6   8   10 m't.

sheet of paper must be applied, and thus the record is continued. Fig. 5 represents the appearance of a sheet containing a day's observations. A graduation upon the vertical side of the sheet indicates differences of height, while a graduation upon the horizontal side indicates the corresponding hours of observation.

17. *Hardy's Self-registering Barometer* is a siphon barometer, A B C, both ends of the tube having the same diameter. Upon the surface of the mercury at C rests an iron float, to which is attached a cord passing over a pulley, P, and from the other end of the cord is suspended a counterpoise, D D. The float is thus made to rise and fall with the mercury in the barometer, without interfering with the free motion of the mercury, and this mo-

Fig. 6.

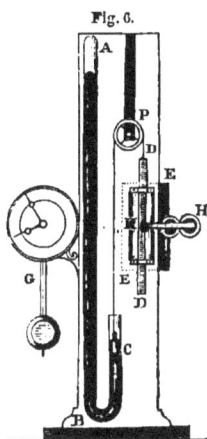

tion is copied by the weight. This weight carries a pencil whose point is very near to a large vertical cylinder, E E, which turns uniformly about its axis. This cylinder, which is covered with a sheet of paper, is made to revolve by means of the clock, G. Every half hour this clock moves a hammer, H K, whose head strikes the weight, D D, by which means the point of the pencil is pressed against the cylinder, and makes a mark whose position indicates the height of the mercury in the barometer. On the sheet of paper is traced a horizontal line divided into equal parts to indicate the hours of the day. The series of points thus marked upon the paper shows the movement of the barometer during 24 hours.

18. *Hough's Printing Barometer.*—Mr. G. W. Hough, Director of the Dudley Observatory at Albany, has invented an instrument which furnishes automatically a printed record of the pressure of the atmosphere for every hour of the day. For this purpose he employs a siphon barometer, and a float resting upon the mercury in the open arm. This float supports a small platinum disk which is placed horizontally between the points of two wires which communicate with a voltaic battery. These wires are supported by a screw, S, which is attached to a toothed wheel, W. When the mercury rises in the short leg of the siphon, the platinum disk is raised, and touches the upper wire, closing the circuit through an electro-magnet, *advancing* the wheel W one tooth, and *raising* the screw S; and so long as the mercury continues to rise, the screw S rises also. When the mercury in the siphon falls, the under side of the platinum disk is brought in contact with the point of the lower wire, closing the circuit through another magnet, moving the wheel W one tooth *backward*, and *depressing* the screw S. Thus the screw S is made to rise or fall with the mercury in the barometer. This screw carries a pencil, which traces upon a revolving cylinder a line showing the minutest movements of the column of mercury during a period of twenty-four hours. This same screw also gives motion to a series of wheels

B

which carry types, by which at the end of every hour the height
of the column of mercury is printed on a slip of paper to the accuracy of the thousandth part of an inch.

.19. *Mean Height of the Barometer.*—If we record the height of
the barometer for each hour of the day, after it is corrected for
temperature and capillarity, and divide the sum of the results by
24, we obtain the *mean height for the day.* If we divide the sum
of the daily means for a month by the number of days, we obtain the *mean height for the month.* If we divide the sum of the
monthly means by 12, we obtain the *mean height for the year.* If
we divide the sum of the annual means for a long period by the
number of years, we obtain the *mean height of the barometer* for
the place of observation. The mean height of the barometer at
Boston is 29.988 inches.

20. *Influence of Latitude.*—The mean height of the barometer
at the level of the sea varies with the latitude of the place. Near

Fig. 7.

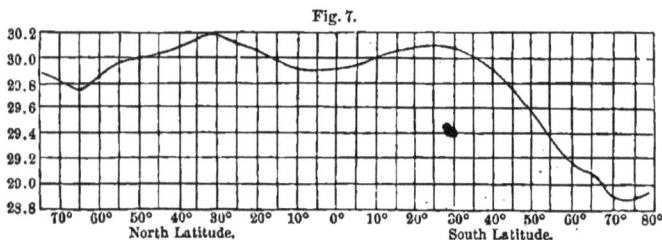

the equator the mean height of the barometer at the level of the
sea is 29.927 English inches. In the northern hemisphere this
pressure increases with the latitude up to 32°, where the mean
height of the barometer is 30.210 inches; the pressure thence
diminishes up to latitude 64°, where the mean height is 29.652
inches; from which point the pressure slightly increases as we
advance northward, being in latitude 78° equal to 29.775 inches.
In the southern hemisphere the barometer is highest near the
parallel of 25°, being there 30.11 inches; the pressure thence diminishes up to latitude 70°, where the mean height is only 28.88
inches, and in latitude 76° the mean pressure is 28.95 inches.
This variation of pressure in different latitudes is shown by Fig.
7. If the atmosphere were at rest, its pressure at the level of the

sea ought to be every where nearly the same. This inequality of pressure must then be due to the *movements* of the atmosphere, as will be explained hereafter, pages 84 and 147.

21. *Inequality of the Monthly Means.*—The mean height of the barometer is not the same for each month of the year, being generally less in summer than in winter. At many places the inequality amounts to half an inch, while at other places it almost entirely disappears. At Pekin, in China, the mean height of the barometer is least in July, from which time the mean pressure increases uninterruptedly to January, after which it declines uninterruptedly to the next July; the pressure in January exceeding that in July by three fourths of an inch. A similar law prevails throughout a considerable portion of the continent of Asia. The cause of this fluctuation will be explained on page 63.

In the middle latitudes of Europe and America, the mean height of the barometer is usually about the same for each month of the

Fig. 8.

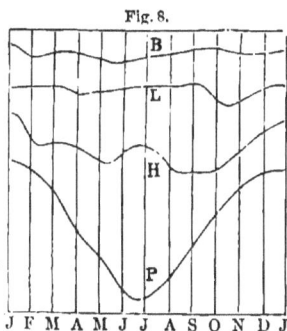

year. At Boston there are no two months whose mean pressures differ by more than one tenth of an inch. A similar remark is applicable to London and Paris. These variations of pressure are conveniently represented by means of curve lines. We draw upon a sheet of paper a horizontal line J J, and divide it into twelve equal parts to represent the different months of the year, and through these points of division we draw a system of vertical lines. Upon each of the vertical lines we set off the mean height of the barometer for the corresponding month, and connect all these points by a broken line. We thus obtain a line whose curvature represents the mean motion of the barometer for each month of the year. The four curves of Fig. 8 show the motion of the barometer: P for Pekin, H for Havana, L for London, and B for Boston. See Table X.

22. *Hourly Variations.*—If we record the height of the barometer for each hour of the day, during a long period of time, and take the mean of all the observations for each hour, we shall find

that these averages are not equal to each other. The height of the barometer is greatest about 10 A.M., and least about 4 P.M. Smaller fluctuations are also observed during the night, the barometer attaining a second maximum about 10 P.M., and a second minimum about 4 A.M. The amount of this diurnal oscillation is greatest at the equator, where its value is 0.104 inch, and it diminishes as we proceed toward either pole. In latitude 40° it is reduced to 0.05 inch; and in latitude 70° it is only 0.003 inch. This oscillation is due partly to changes in the pressure of the gaseous atmosphere, and partly to changes in the amount of vapor present in the air, as will be shown on page 62.

These variations of pressure may be represented by curve lines. We draw upon a sheet of paper several vertical and equidistant

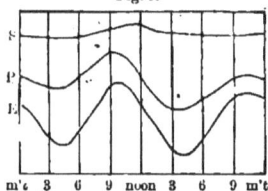

Fig. 9.

lines to represent the hours of the day. We set off upon each of the vertical lines the mean height of the barometer for the corresponding hour, and connect all these points by a broken line. We thus obtain a line whose curvature represents the mean motion of the barometer for each hour of the day. The three curves of Fig. 9 show the motion of the barometer, E for the equator, P for Philadelphia, and S for St. Petersburg. These curves are seen to have two daily maxima and two daily minima. See Table XI.

23. *Inequality depending on the Position of the Moon.*—There is a small fluctuation in the pressure of the atmosphere depending on the position of the moon; but this variation is exceedingly minute, and can only be detected by taking the mean of the most accurate observations continued for a long period of time. At Singapore, latitude 1° 18', when the moon is on the meridian, the pressure of the atmosphere is 0.0057 inch greater than when the moon is six hours from the meridian. At St. Helena, latitude 15° 55', when the moon is on the meridian, the pressure of the atmosphere is 0.004 inch greater than when the moon is six hours from the meridian. In higher latitudes the difference of pressure is still less. These results indicate a feeble tide in our atmosphere, similar to the tides of the ocean.

24. *Accidental Variations.*—The non-periodic oscillations of the barometer are far greater than the periodical ones.   In the middle latitudes the barometer is almost constantly in motion, and the fluctuations are so great and so irregular that the periodical movements are only detected by taking the mean of a long series of observations.   The difference between the greatest and least heights of the barometer during a single month is called the *monthly oscillation ;* and by combining observations extending over a great number of years, we obtain the *mean monthly oscillation.*   The mean monthly oscillation is least in the neighborhood of the equator, and increases as we approach the poles.   At the equator it is but little over one tenth of an inch ; in latitude 30° it is four tenths of an inch ; in latitude 45°, over the Atlantic Ocean, it is one inch ; in latitude 65° it is one inch and a third ; and in latitude 78° it is one inch and a fifth.   During the three winter months, the mean monthly oscillation is about one third greater than the numbers here stated.   Over the continents of Europe and America, the oscillations are generally less than over the Atlantic on the same parallel.

25. *Extreme Fluctuations of the Barometer.*—The extreme fluctuations of the barometer are much greater than the numbers here given.   The greatest height which the barometer at Boston has attained in 37 years is 31.125 inches, and the least is 28.47 inches ; the difference being 2.655 inches, or $\frac{1}{11}$th of the average height of the column.   At London, the greatest observed range of the barometer is three inches, while at St. Petersburg and in Iceland it is 3.5 inches.   At Christiansborg, near the equator, the entire range of the barometer in five years was 0.47 inch.

26. *Influence of the Wind.*—The height of the barometer is sensibly influenced by the direction of the wind.   At Philadelphia the barometer generally stands highest when the wind is northeast, and lowest when the wind is west or southwest, the mean difference in the height of the barometer for these different winds being a quarter of an inch.   Throughout the northwest part of Europe the barometer stands highest when the wind is northeast, and lowest when the wind is south, the mean difference for these two winds being 0.22 inch.

27. *Pressure affected by Height of Station.* — When a barometer is elevated above the surface of the earth, the column of mercury sinks, because the force which sustains the column, that is, the weight of the superincumbent air, is diminished. By comparing the height of the mercury in barometers at two stations, one of which is above the other, we ascertain the weight of a column of air extending from the lower to the higher station. For example, if the mercury in the lower barometer stands at 30 inches, and in the higher barometer at 29 inches, it follows that a column of air extending from the lower to the higher station has a weight equal to that of a column of mercury one inch high. Now the density of mercury is 10,464 times that of air; hence a fall of one inch in the barometer would indicate an elevation of 10,464 inches, or 872 feet, above the first station, provided the density of the air were the same at both stations.

28. *Heights measured by Barometer.* — Since the air is readily compressed, its density rapidly diminishes as the height increases. Mathematicians have endeavored to discover the exact relation between the change of barometric heights and the difference of level of the two stations of observation. Laplace deduced a formula which is designed to take account of all the corrections required for attaining the greatest accuracy, such as the change of temperature of the air between the two stations, the diminution of gravity on a vertical line, etc. According to this formula, in the neighborhood of New York, when the atmosphere is at its mean state, if we ascend above the level of the sea

917 feet, the barometer sinks 1 inch.
1860 " " " 2 inches.
2830 " " " 3 "
3830 " " " 4 "
4861 " " " 5 "

Table IX, page 260, affords the means of determining the difference in the heights of any two places by means of barometric observations.

CHAPTER II.

TEMPERATURE OF THE AIR AND OF THE EARTH.

29. *Climatology.*—Climatology is the science of climates. By the climate of a country we understand its condition relative to all those atmospheric phenomena which influence organized beings. Climate depends upon the mean temperature of the year; upon that of each month and each day; upon the maximum and minimum temperatures; upon the frequency and suddenness of the atmospheric changes; upon the transparency of the atmosphere and the amount of solar radiation; upon the moisture of the air and the earth; upon the prevalence of fogs and dew; the amount of rain and snow; the frequency of thunder-storms and hail; the direction, force, and dryness of the winds, etc. All these particulars can only be determined by long and careful observations.

30. *Thermometer.*—The changes of temperature of the air are measured by means of the thermometer. This instrument generally consists of a small glass bulb, to which is attached a long glass tube, having a very small bore. The bulb is filled with mercury or alcohol, which also rises somewhat within the tube. Now mercury and alcohol are very much expanded by an increase of heat, while glass expands very little. If, then, the temperature of the thermometer rises, since the mercury expands more than the bulb which contains it, it overflows the bulb, and is forced up into the small tube. If the temperature falls, the mercury contracts more than the glass bulb, and the mercury in the tube descends to fill the vacuum created in the bulb. Thus the changes of temperature to which the thermometer is subjected are indicated by the ascent or descent of the mercury in the small tube.

31. *Graduation of the Scale.*—In order that we may have an intelligible measure of these changes of temperature, the tube must be graduated according to some general principles. We need two invariable temperatures for the determination of two fixed points upon the scale. The temperatures generally adopted for this purpose are those of melting ice and boiling water; and the interval between these points is variously divided in different countries. Upon Fahrenheit's thermometer, melting ice is marked 32°, and boiling water 212°, the interval being divided into 180 equal parts. The same graduation is extended downward from 32° to zero, and may be continued below zero as far as is desired. Degrees below zero are distinguished by the minus sign. Thus we may have a temperature of 40° above zero, or 40° below zero. Fahrenheit's scale is generally used in England and the United States.

Upon the Centigrade thermometer, the freezing point is marked 0, and the boiling point 100. This thermometer is generally used in France. Upon Reaumur's thermometer, the freezing point is marked 0, and the boiling point 80. This thermometer is generally used in Germany and Russia.

32. *Requisites of a good Thermometer.*—It is evident that the degrees upon the thermometer scale should correspond to equal volumes of mercury. If the tube of the thermometer were throughout of uniform bore, then the divisions upon the scale should be throughout of equal length; but if the tube be not of uniform bore, these equal volumes will correspond to unequal lengths upon different parts of the scale. Now it is impossible to obtain a glass tube perfectly cylindrical, and therefore when an accurate graduation is required, we proceed as follows: Having selected a tube whose bore is as nearly uniform as possible, we introduce

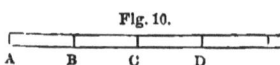

Fig. 10.

into it a short column of mercury A B, and mark its extremities upon the tube. Then, by agitating the tube, we push the mercury along to B C, so that its left extremity may occupy the same position as the right extremity in the first trial; and mark the extremity C. The volumes of A B and B C are evidently equal. We thus crowd the column of mercury along from one end of the tube to the other, and divide it into portions of equal volume. Each of these portions, A B, B C, C D, etc., should then be made to contain the same number of divisions of the scale.

When a standard thermometer has been constructed in this manner, other thermometers are frequently graduated by comparison with it at several different points of the scale.

33. *Self-registering Thermometers.*—It is frequently desired to determine the greatest heat or the greatest cold experienced during a day or some longer interval of time.   To do this with an ordinary thermometer, it is necessary that the instrument be frequently observed at short intervals.   Such observations are very laborious; and in order to diminish this labor, *self-registering* thermometers have been invented.

The following is one form of a thermometer for registering the *highest* temperature.   A small piece of steel wire *c*, about half an inch in length, and finer than the bore of the thermometer, is introduced into the tube of a mercurial thermometer above the mercury.   The thermometer is placed with its stem A B in a

Fig. 11.

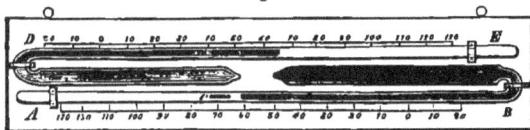

horizontal position, and the steel index is brought into contact with the extremity of the column of mercury.   Now, when the heat increases and the mercury expands, the index *c* will be thrust forward; but when the temperature falls, and the mercury contracts, the index will be left behind.   The point of the scale where the index is found shows therefore the *greatest* degree of heat to which the instrument has been subjected since the last observation.

34. *Minimum Thermometer.*—The *lowest* degree to which the thermometer has fallen may be indicated as follows: A spirit thermometer is placed with its stem D E horizontal, and within the tube is a very fine glass rod, or index, *n*, about half an inch in length, and a little smaller than the bore of the tube.   This index is immersed in the column of alcohol, but must be brought into contact with the extremity of the column.   On account of the capillary adhesion between the alcohol and the glass, when the alcohol contracts, it drags along with it the glass index; but when the alcohol expands, it passes by the index without displacing it, so that the position of the index shows the *lowest* temperature to which the instrument has been subjected since the last observation.

These instruments are especially adapted to indicate the maximum and the minimum temperature in twenty-four hours. The steel index being placed in contact with the mercury, and one extremity of the glass index being made to coincide with the extremity of the column of alcohol, the position of the two indices on the following day will show what has been the highest and what has been the lowest temperature during the last twenty-four hours.

35. *Phillips's Maximum Thermometer.*—In this thermometer a small portion of the column of mercury is separated from the remainder of the column by an extremely minute speck of air, so that this detached column serves the same purpose as the steel wire in the ordinary maximum thermometer. The end of this detached column remains at the point of maximum temperature, while the other part of the column retreats toward the bulb when the temperature declines. By bringing the instrument to a vertical position with the bulb downward, the detached portion descends *nearly* into contact with the remainder of the column, but is prevented from uniting with it by the presence of the air speck. This instrument is susceptible of very great precision.

36. *Photographic Register of the Thermometer.*—In some observatories, the height of the thermometer is registered photographically, in a manner similar to that described in Art. 16. The light of a lamp is concentrated by means of a lens, so as to strike upon the summit of the column of mercury in the thermometer. A sheet of paper suitably prepared for photographic experiments is placed behind the thermometer, and receives the shadow cast by the mercury. By means of clock-work, the paper is carried steadily forward, and thus the column of mercury leaves upon the paper a record of its height at each instant of the twenty-four hours. This is in some respects the best self-registering thermometer known, although the record is usually not very sharp, and therefore not as accurate as could be desired.

37. *Cause of the variations of Temperature.*—The sun is the principal cause of the variations of the temperature of the atmosphere. The amount of heat which the sun communicates in a given time depends upon the elevation of the sun above the horizon, and upon the transparency of the atmosphere. The difference between summer and winter depends upon the time that the sun remains above the horizon, as well as upon its distance from the zenith of the observer.

38. *How the Atmosphere is Heated.*—The atmosphere is heated in three ways: by the direct rays of the sun; by contact with the warmer earth; and by the radiation and reflection of heat from the earth.

A portion of the rays of heat which are emitted by the sun are absorbed by our atmosphere before they can reach the earth's surface. It is estimated that in clear weather the atmosphere absorbs about one fourth of the rays which traverse the atmosphere vertically. The remaining rays are received upon the earth's surface, by which means the earth is heated. This heat is thence communicated to the air which rests upon the earth; and this air, being thereby rendered lighter, rises and gives place to colder air from above. This in turn, by contact with the earth, becomes heated, and rises, and thus there is maintained a continued circulation between the strata of air in the neighborhood of the earth.

A portion of the heat which the earth receives from the sun radiates into space. These rays are partly absorbed by the air, especially by its lower strata, and these strata, in their turn, diffuse invisible rays of heat in every direction.

The effect of the direct rays of the sun is plainly seen in winter when the ground is covered with snow. In the vicinity of trees and posts the snow disappears more rapidly than it does where the surface of the snow is entirely unbroken. This is because the rays of the sun are absorbed by the dark surface of the trees more readily than they are by the snow. Thus the trees are warmed, and these, in their turn, send out rays of heat by which the adjacent snow is melted.

39. *Proper Exposure of a Thermometer.*—For the purpose of measuring the temperature of the air, a thermometer should be exposed in the open air, where the circulation is unobstructed. It should face the north, and should be always in the shade. It should be removed at least a foot from the wall of the building, and should be elevated about ten feet from the ground. It should be protected against the heat reflected by neighboring objects, such as buildings or a sandy soil, and it should be sheltered from the rain. If the thermometer should happen to become moistened by rain, the bulb should be carefully dried about five minutes before making the observation; since drops of water, by their evaporation, would lower the temperature of the mercury in the bulb.

In order to secure all these advantages, it is generally found necessary to cover the thermometer with a wooden frame of open lattice-work; but this covering should be such as to allow a perfectly free circulation of air about the thermometer, and it should be such as readily to acquire the temperature of the surrounding air.

Fig. 12.

Fig. 12 represents a frame adopted at Greenwich Observatory for supporting the thermometers. It consists of two parallel inclined boards, with a small projecting roof, beneath which the thermometers are suspended, so that the air circulates freely about the bulbs. The whole frame revolves on an upright post, and the inclined side is always turned toward the sun.

40. *Hourly Observations of the Thermometer.*—In order to determine the laws which govern the variations of the temperature of the atmosphere, we require that observations should be made from hour to hour, both night and day, throughout a period of several years. Such observations have been made at many different places. The most extensive series of this kind in North America was made at Toronto, where bi-hourly observations were continued for ten years. At Philadelphia, hourly observations were made for two and a half years, and bi-hourly observations for another two and a half years. At Washington, observations every two hours were continued for two and a half years. Similar observations upon a less extensive scale have been made at a few other places in this country.

41. *Hourly Variations of Temperature.*—The temperature of a place changes from one hour to another, according to the distance of the sun from the horizon. If we take the average of all the temperatures observed at each hour of the day for a long period of time, we shall find that the mean hourly variations of temperature are extremely regular. Figure 13 shows the general law of the change of temperature at New Haven. The abscissas represent the hours of the day, and the ordinates the temperatures observed.

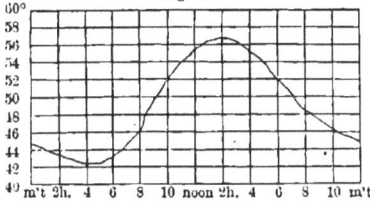

Fig. 13.

We see that on each day there is a maximum and minimum of temperature. At New Haven, the minimum occurs about an hour before the rising of the sun, and the maximum about two hours after noon. In the average of the entire year, the temperature is increasing during nine hours of the day, and decreasing during the remaining fifteen hours of the day.

The highest temperature of the day should occur when the amount of heat lost each instant by radiation is just equal to the heat received from the sun. Before midday, the earth receives from the sun more heat than it loses by radiation, and its temperature rises. After noon, the earth receives each instant from the sun less heat than it did at noon; but the heat received is still greater than that which is lost by radiation. Hence the maximum takes place some time after noon. During the night we receive no direct heat from the sun, and the earth cools by radiation. The lowest temperature should occur when the heat received each instant from the returning sun is just equal to the loss by radiation. This occurs about an hour before sunrise.

42. *Mean Temperature of a Day.*—The mean temperature of a day is the mean of the twenty-four observations taken at each hour of the day. Since hourly observations of the thermometer are very laborious, it is important to discover simpler methods of ascertaining the mean daily temperature. The following are the principal methods which have been employed for this purpose.

43. *From the Maximum and Minimum Temperatures.* — The

mean of the highest and lowest degrees of the thermometer during the twenty-four hours differs but little from the mean derived from hourly observations; and when self-registering thermometers can be procured, this is a convenient mode of obtaining the mean daily temperature. This method is not, however, entirely accurate, since the mean of the two daily extremes is generally a little greater than the mean for the twenty-four hours. At New Haven the average difference of the two results for the entire year is about half a degree; being nearly an entire degree in winter, and about zero in summer. When the highest accuracy is required, a small correction should therefore be applied for the error of this method.

44. *From Observations at a single Hour.*—When self-registering thermometers can not be obtained, one of the following methods may be practiced: Twice during each day the height of the thermometer must coincide with the mean temperature of the day. At New Haven, this coincidence occurs about a quarter before nine in the morning, and also about a quarter before eight in the evening. We should, therefore, obtain very nearly the mean temperature by a single daily observation at either of these hours. Since, however, at these times, the changes of temperature are quite rapid, a considerable error would result if the observation were made a little too soon or a little too late. Moreover, these hours vary at different localities, and they also vary with the season of the year, so that it is better to deduce the mean temperature from two or more daily observations.

45. *From two Hours of the same Name.*—It is found that the mean temperature of any two hours of the same name differs but little from the mean of the twenty-four hours. Thus the mean of two observations at 6 A.M. and 6 P.M., is nearly the same as that of two observations at 7 A.M. and 7 P.M., or 8 A.M. and 8 P.M., etc.; and at New Haven the mean of two observations at 10 A.M. and 10 P.M. differs only about one third of a degree from the mean of the twenty-four hours. These hours (10 A.M. and P.M.) are better than any other two hours for furnishing the mean temperature; and the mean of these hours is generally nearer the mean temperature of the day than the mean of the two daily extremes.

46. *From three Daily Observations.*—A still more reliable result may be derived from three daily observations. The mean of observations at 6 A.M., 2 and 9 P.M., gives very nearly the mean temperature of the day. The mean of observations at 7 A.M., 2 and 9 P.M., is a little too great; but if we add twice the nine o'clock observation to the sum of the other two observations, and divide the result by four, the error of the result for the separate months at New Haven in only one instance exceeds a quarter of a degree, and for the entire year differs but one hundredth of a degree from the true mean temperature. It is found that for nearly every variety of climate this method furnishes the best result which can be deduced from any three daily observations, and these are therefore the three hours to be generally recommended to observers. See Tables XV. and XVI.

47. *Mean Temperature of the Months.*—The mean temperature of a month is found by dividing the sum of the daily means by the number of days. Figure 14 shows the mean temperature of each month of the year at New Haven, and also the mean maximum and minimum for the month, according to 86 years of observations. The months are arranged upon the horizontal line,

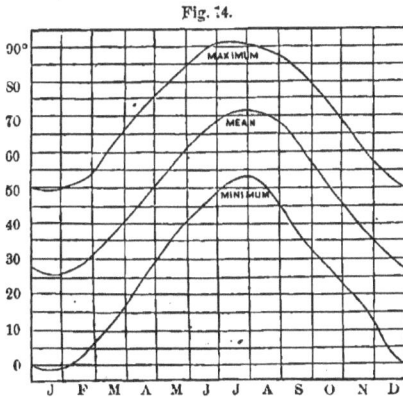

Fig. 14.

and the temperature for each month is represented by the corresponding ordinate. The upper and lower curves pass through the maxima and minima temperatures for the different months, and the intermediate curve corresponds to the monthly mean temperature.

We find that at New Haven, 1st. The warmest months of the year are July and August, and the maximum for the year occurs near July 24th. 2d. The coldest month of the year is January, and the minimum for the year occurs near January 21st. 3d. The difference between the minimum and maximum for each month is greater in the cold months than in the warm months. 4th. The mean temperature of the month of April is two degrees *below* the mean temperature of the year, while that of October is two degrees *above* the mean for the year; and the mean temperature of the two months April and October differs less than one tenth of a degree from the mean temperature of the year.

48. *Monthly Change in different Latitudes.*—At most places in the northern hemisphere, the change of temperature for the different months follows a law similar to that above described for New Haven. We find at most places that the average heat goes on increasing from day to day, uninterruptedly from March until some time in summer, and after that time the mean heat of each day decreases uninterruptedly until some time in winter. The time, however, of the annual maximum and minimum varies with the latitude of the observer. Near the equator, the entire annual variation of temperature is very small, and the greatest cold may occur in any month from November to March, or even from July to September. Indeed, at some places near the equator there are two annual maxima of temperature and two annual minima. But in the extreme southern part of the United States, the greatest cold usually occurs in December; near the parallel of 40°, it occurs about the middle of January; in the northern part of the United States, about the first of February; at Toronto it occurs as late as the middle of February; and in latitude 78°, the greatest cold occurs in March.

Throughout most of the United States, the maximum temperature occurs about the middle of July; but at some places north of the United States, the maximum does not occur until the 10th of August. See Table XVII.

49. *Cause of these Peculiarities.*—If the temperature at any place depended simply upon the direct momentary influence of the sun, the maximum would coincide with the summer solstice; but during the most of summer the heat received from the sun during the day is greater than the loss of radiation during the night, and the maximum occurs when the loss by night is just equal to the gain by day. During the autumn, the loss by night is much greater than the gain by day, and the mean temperature rapidly falls. The minimum occurs when the gain by day is just equal to the loss by night, and this generally takes place some time after the winter solstice.

The time of maximum or minimum temperature depends not simply upon the sun's altitude at noon, but also upon the number of hours during which the sun is above the horizon ; that is, upon the relative length of the days and nights. The minimum occurs later in high than in low latitudes on account of the shortness of the winter days in high latitudes; and the maximum occurs later on account of the greater length of the summer days in high latitudes.

50. *Mean Temperature of a Place.*—The mean temperature of a year is found by taking the average of all the monthly temperatures for the year. This annual mean is not the same every year at the same place; nevertheless, the difference between the coldest and hottest years seldom exceeds ten degrees. At New Haven the hottest year which has occurred in a period of 86 years was that of 1828, and the coldest year was that of 1836, the extreme range of the annual temperature in 86 years being 6°.3.

At Breslau, in Prussia, the extreme range of the annual temperature in 66 years has been ten degrees.

By taking the average of the mean annual temperatures for a great number of years, we obtain the mean temperature of a place. To determine this mean temperature with considerable accuracy for a variable climate, observations should be continued at least a quarter of a century, in order that the accidental differences between successive years may compensate each other.

This mean temperature of a place is sensibly constant from one century to another, and there is no sufficient reason for believing that the mean temperature of any place on the earth's surface has changed appreciably in two thousand years.

51. *Non-periodic Variations.*—Besides the periodic variations of temperature, there are accidental variations due to causes which will be mentioned hereafter. These fluctuations of temperature are frequently experienced simultaneously over large portions of the globe; and we frequently find that at the same time in other parts of the world changes of temperature are observed in the opposite direction.

C

DISTRIBUTION OF HEAT OVER THE EARTH'S SURFACE.

52. *Temperature of different Latitudes.*—If we follow a meridian from the equator toward either pole, we shall find that the mean temperature generally decreases, but not uniformly. On the contrary, there are places where, as we proceed toward the pole, the mean temperature rises instead of falling. These irregularities are due to local causes, which vary upon different meridians, so that the points of equal mean temperature are not situated upon a parallel of latitude.

53. *Isothermal Lines.*—In order to represent all the observations of temperature conveniently upon a map, we draw a line connecting all those places whose mean temperature is the same. Such a line is called an *isothermal* line. In the neighborhood of the equator, the mean annual temperature is usually about 80°. In Africa and the Indian Archipelago, the mean temperature near the equator is about 82°, and in a few localities it is still higher. In a few places the mean temperature of a single year has been known to rise to 85°, and even higher. The area having a *mean temperature of* 80° and upward, forms a belt of over 1000 miles in breadth for more than half the circumference of the globe; for about a quarter of the circumference this belt has a breadth varying from 1000 miles to zero; and for thirty or forty degrees of longitude, the mean temperature near the equator does not exceed 79°. See Table XVIII.

The *isothermal line of* 70° is a line gently undulating, but generally is nearly parallel to the equator. In the northern hemisphere, this line passes through Galveston, New Orleans, Mobile, and St. Augustine; through the Island of Teneriffe; through Alexandria, in Egypt; and Canton, in China.

The *isothermal line of* 60° passes through Sacramento, California; Memphis, Tennessee; Chapel Hill, North Carolina; Norfolk, Virginia; through the northern part of Spain; Rome, in Italy; a little south of Constantinople; near the south end of the Caspian Sea; and through Shanghai, in China.

The *isothermal line of* 50° passes through Puget's Sound, on the Oregon coast; through Burlington, Iowa; Pittsburg, Pennsylvania; New Haven, Connecticut; Dublin, in Ireland; Brussels, in Belgium; and Vienna, in Austria; near the northern shore of the Caspian Sea; and a little north of Pekin, in China.

The *isothermal line of* 40° passes through the middle of Lake

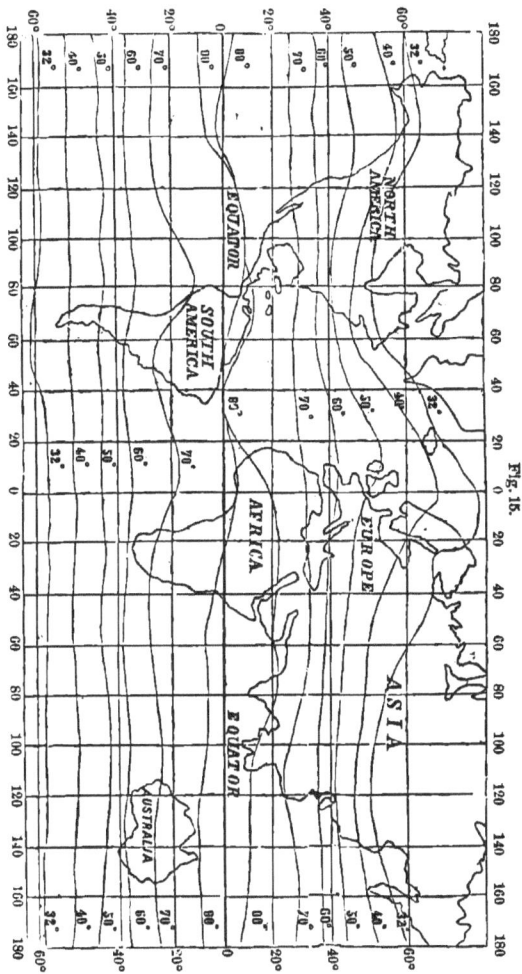

Superior; through Hanover, New Hampshire; through Quebec; a little south of Iceland; through Upsala, in Sweden; through Petersburg and Moscow, in Russia.

The *isothermal line of* 32° is an undulating oval curve, whose centre is near the north pole, and which is elongated in the direction of the continents of America and Asia. Over the conti-

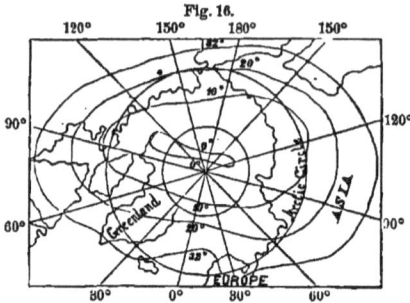

nents this line descends to latitude 52°, but on the coast of Norway it rises as high as latitude 72°. The longer diameter of this curve is nearly twice that of its shorter. This line passes a little south of Behring's Straits; near the northern shore of Lake Superior; through the south margin of James's Bay; the southern part of Greenland; a little north of Iceland; through North Cape, in Norway; and through Barnaul, in Siberia. Throughout the entire area inclosed by this line the mean annual temperature is below that of melting ice. All these lines are represented on figures 15 and 16.

It is not claimed that all of these lines have been traced by actual observation, and the positions assigned them are liable to some degree of uncertainty; but the observations are so numerous and so well distributed as to leave little doubt respecting the approximate position of the isothermal lines. ·

54. *Mean Temperature of the North Pole.*—At several places in the Arctic regions the mean temperature has been found to be but little above zero; and at Van Rensselaer Harbor, in latitude 78°, the mean temperature is two and a half degrees below zero. It is probable that near the north pole there is a considerable area whose temperature is below zero of Fahrenheit. From the form of the neighboring isothermal lines we conclude that this area is an oval nearly 2000 miles in length and 700 miles in breadth, and it lies chiefly on the American side of the north pole. It is even doubtful whether the north pole is at all included in this area. The coldest spot in the northern hemisphere appears to be north of the American continent in latitude 80° to 85°, and its mean temperature is probably at least five degrees below zero. See Table XIX.

55. *Two Sides of the Atlantic compared.*—The mean temperature of the eastern side of the Atlantic is much warmer than that of the western, upon the same parallel of latitude. The mean temperature of New York is about the same as that of Dublin, although Dublin is 13° north of New York. Near Lake Superior, in latitude 50°, we find the same mean temperature as at the North Cape, in latitude 72°.

This high temperature of the European coast is due to the high temperature of the North Atlantic, combined with the prevalent westerly winds. By means of the Gulf Stream, the waters of the equatorial regions are conveyed into the North Atlantic, and a portion of this warm current extends northward between Iceland and the British Islands, and continues to the Arctic Ocean. The temperature of the North Atlantic is thus raised much above what is due to its latitude; and since throughout the middle latitudes the prevalent winds are from the west, this heat of the ocean is communicated to places on the eastern side of the Atlantic, but not to those on the western side.

56. *Two Sides of the Pacific Ocean.*—The currents of the Pacific Ocean produce an effect similar to the currents of the Atlantic, and there is a corresponding difference between the temperatures of places on opposite sides of the Pacific Ocean, and consequently a marked difference between the temperatures of places on the Atlantic and Pacific coast of North America, although situated on the same parallel of latitude. The isothermal line of 50° is found ten degrees of latitude farther north on the Pacific coast than it is on the Atlantic coast. Sitka, in latitude 57° 3′, has about the same mean temperature as Eastport, Maine, in latitude 44° 54′.

57. *Northern and Southern Hemispheres compared.*—The mean temperature of the northern hemisphere is sensibly higher than that of the southern.

On the parallel of 10°		
" " 20°	there is an average dif-	2° .1.
" " 30°	ference amounting to	3° .4.
" " 40°		2° .9.
		1° .9.

We have not sufficient observations to decide whether this difference continues in the higher latitudes.

This difference in the temperature of the two hemispheres probably results from the unequal distribution of land and water. The northern hemisphere contains much more land than the southern. In the southern hemisphere, the sun's rays fall chiefly upon the water, and are employed in converting water into vapor, in which process a large amount of heat is rendered latent. This heat again becomes sensible when this vapor is condensed in rain. But rain is much more frequent in the northern hemisphere than in the southern. From a comparison of records, embracing in the aggregate a period of nearly one thousand years of observations, it appears that the number of rainy days in the North Atlantic is fifty per cent. greater than it is in the South Atlantic. Thus we find that the southern hemisphere is cooled by evaporation more than the northern, and the northern is warmed by the condensation of vapor more than is the southern, by which means the average temperature of the northern hemisphere is rendered sensibly higher than that of the southern.

58. *Hottest and coldest Months compared.*—The climate and productions of a country are very imperfectly indicated by its *mean* temperature. Two places may have the same mean temperature, yet differ greatly in their extreme temperatures, and consequently also in their vegetable productions. Thus the mean temperature of New York is the same as that of Liverpool; yet the difference between the mean temperature of the three summer months and that of the three winter months is twice as great in New York as it is in Liverpool. Throughout England the heat of summer is insufficient to ripen Indian corn; while the ivy, which grows luxuriantly in England, can scarcely survive the severe winters of New York.

There are some places where the mean temperature of the hottest month of the year differs less than five degrees from that of the coldest month. This is true of some of the West India Islands, and also in the Indian Archipelago. At Singapore, the mean temperature of January differs but $3\frac{1}{2}°$ from that of July.

On the contrary, there are some places where the mean temperature of the hottest month differs 50°, 80°, and even 100° from that of the coldest month. At Quebec, this difference amounts to 60°; at Fort Churchill, on Hudson's Bay, the difference is 86°; and at some places in Siberia the mean temperature of January is more than 100° below that of July. See Tables XX. and XXI

59. *Climates either Marine or Continental.*—The most uniform temperature is found to prevail upon islands, while the greatest range of temperature prevails in the interior of continents. Hence climates may be characterized as either *marine* or *continental.* The temperature of the ocean varies but little from summer to winter, while that of the land may vary more than 100°. Hence those places whose temperature is mainly controlled by the ocean have an equable climate, while those which are but little affected by the ocean have an extreme climate.

The annual range of temperature is much less on the eastern than on the western side of the Atlantic, because the prevalent winds are from the west. Hence, on the western coast of the Atlantic, where the prevalent winds come from the land, the climate is essentially continental, but upon the eastern side of the Atlantic, where the prevalent winds come from the sea, the climate is mainly controlled by the ocean.

60. *Highest observed Temperature.*—Although the highest *mean* temperature is found near the equator, yet the thermometer frequently rises higher in the middle latitudes than it does at many places under the equator. Thus, at Singapore, under the equator, the thermometer never rises above 95°, while at New York and at Paris the thermometer has been known to rise to 104°. At Mosul, in Armenia, the thermometer has been known to rise to 117°; at Fort Miller, California, to 121°; in India, to 132°; and on the Great Desert of Africa, to 133°.

These numbers are supposed to indicate the temperature of the air where it circulates most freely. A thermometer exposed to the direct rays of the sun often rises much higher than the preceding numbers. In India, a thermometer whose bulb was covered with black wool rose in the sun to 164°; and a thermometer placed inside of a blackened box, covered with glass, has been known to rise to 248°.

61. *Lowest observed Temperature.*—The lowest temperatures any where observed have been in North America and Siberia. The lowest temperature observed at Singapore is 66°; at Key West, 45°; at Paris and New York, —10°; at New Haven, —24°; and at Montreal, —38°. At New Lebanon, New York, at Franconia, New Hampshire, and at several places in New England, mercury froze in January, 1835, indicating a temperature of 40° below zero. Dr. Kane, in latitude 78°, observed a temperature of 67° below zero; and Captain Back, at Fort Reliance, in latitude 62°, observed a temperature of 70° below zero; while in Siberia the thermometer has been known to fall to 76° below zero.

62. *Range of Temperature.*—By combining these results, we find that at Singapore the entire range of the thermometer is only 29°, while at New York it is 114°; at Montreal it is 140°, and at Fort Reliance, in latitude 62°, the thermometer in four months varied from −70° to +81, being a range of 151°.

The entire range of the temperature of the air any where observed is from −76 to +133°, or 209°.

The range of temperature for a single day in the middle latitudes is often greater than for a whole year in the equatorial regions. At Hanover, New Hampshire, February 7, 1861, at noon, the thermometer stood at 40°; the next morning it stood at −32°, making a range of 72° in 18 hours. See Tables XXII. and XXIII.

TEMPERATURE OF THE AIR AT DIFFERENT HEIGHTS.

63. *Change of Temperature with Elevation.*—As we ascend above the surface of the earth the mean temperature of the air declines. This depression of temperature is observed when we ascend a mountain or rise in a balloon. The rate of decrease varies with the latitude of the place, with the season of the year, as well as the hour of the day. It is more rapid in warm countries than in cold countries, and is most rapid during the hottest months. It is most rapid about 5 P.M., and least rapid about sunrise.

The change is also most rapid near the earth's surface, and diminishes as we ascend. From a long series of balloon ascents, made under the direction of the British Scientific Association, the following results have been obtained for the vicinity of London:

Elevation.	When the Sky is clear.	When the Sky is cloudy.
From 0 ft. to 5000 ft., the decrease is 1° for 239 ft. elev'n.	1° for 271 ft. elev'n.
" 5000 " 10000 " " " 394 "	" 394 "
" 10000 " 15000 " " " 490 "	" 459 "
" 15000 " 20000 " " " 581 "	" 725 "
" 20000 " 25000 " " " 877 "	" 1111 "
" 25000 " 30000 " " " 1190 "	

64. *Cause of this Decrease of Temperature.*—This decrease of temperature as we rise above the earth's surface is mainly due to the expansion of the air. The lower strata of the air, being heated by the sun (Art. 38) and expanded, tend to rise in consequence of their diminished specific gravity. As the air ascends it is subject to a diminished pressure and expands; its heat is diffused through a greater amount of space, by which means a part of its sensible heat becomes latent.

This principle may be proved experimentally by placing a thermometer under the receiver of an air-pump and rapidly exhausting the air, when the thermometer indicates a diminution of sensible temperature. Upon readmitting the air the thermometer rises to its former height.

The atmosphere would be in a condition of equilibrium if a pound of air at all elevations, whether on the summit of a mountain or at the level of the sea, contained the same amount of heat. The atmosphere is perpetually seeking to attain to this condition of equilibrium, but since the sun perpetually acts as a disturbing force, such an equilibrium is never fully attained.

65. *Law of decrease of Temperature with Height.*—We see from the observations of Art. 63 that the diminution of temperature is not proportional to the height; but we find that the temperature is intimately connected with the pressure, as shown by the barometer. The following table presents a summary of these observations for a clear sky:

Barometer.	Temperature.	Difference.	Barometer.	Temperature.	Difference.
10 inches.	$-10°.9$		20 inches.	$15°.3$	
12 "	$-6 .1$	$4°.8$	22 "	$21 .0$	$5°.7$
14 "	$-1 .7$	$4 .4$	24 "	$26 .8$	$5 .8$
16 "	$+3 .7$	$5 .4$	26 "	$32 .7$	$5 .9$
18 "	$+9 .5$	$5 .8$	28 "	$39 .9$	$7 .2$
20 "	$+15 .3$	$5 .8$	30 "	$50 .0$	$10 .1$

Column first shows the pressure indicated by the barometer, and column second the corresponding temperature when the temperature at the earth's surface was 50°. The third column shows the change of temperature corresponding to a change of two inches in the pressure. These differences are greatest near the earth's surface, but after rising one mile they become nearly constant; that is, *the fall of the thermometer is nearly proportional to the fall of the barometer*, the change of the thermometer being about five degrees for a change of pressure amounting to two inches.

Fig. 17.

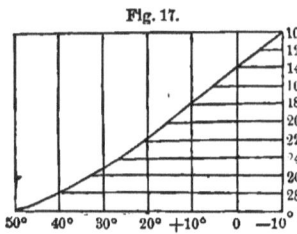

The curve in Fig. 17 shows more readily how the temperature depends upon the pressure. The abscissas represent the observed temperatures from 50° to —11°, and the ordinates show the corresponding pressures from thirty inches to ten inches.

66. *Limit of Perpetual Snow.*—In consequence of this decrease of temperature, the summits of high mountains, even within the tropics, are always covered with snow. The limit of perpetual snow is not the line whose mean temperature is 32°. The snow-line is determined more by the mean temperature of the hottest month than by the mean temperature of the year.

The limit of perpetual snow generally descends as we proceed from the equator toward the poles, but there are many exceptions to this rule. The height of the snow-line depends upon a variety of circumstances: not only upon the mean temperature, but upon the extreme heat of summer; upon the amount of the annual fall of snow; upon the prevalent winds; and upon the proximity of mountain peaks or extensive plains. Under the equator the height of the snow-line varies from 15,000 to 16,000 feet, where the mean annual temperature is 35°. On the Alps, the average height of the snow-line is 8800 feet, where the mean annual temperature is 25°; while on the coast of Norway its height is only 2400 feet, where the mean annual temperature is 21°. Fig. 18 shows the snow-line on several mountains in different latitudes.

Fig. 18.

Numbers 1, 2, and 3 are the Illimani, Aconcagua, and Chimbo-

razo, in South America; 4, 5, and 6 are the Choomalari, Dhaula-
giri, and Caucasus, in Asia; number 7 is the Pyrenees, and 8 the
Alps; number 9 the Sulitelma, in Norway; and number 10 the
island Mageroe. See Table XXIV.

67. *Temperature of the Interplanetary Spaces.*—The temperature
of the air does not continue to sink indefinitely as we rise above
the earth's surface. Its mean temperature can nowhere fall below the temperature of the interplanetary spaces. The space in
which the planets move has a temperature of its own, due to the
radiation of heat from the stars, each of which is a hot body like
our sun. This temperature of space is necessarily lower than the
mean temperature of the polar regions of the earth, for during six
months of the year these are illumined by the sun, from which
they derive a large amount of heat.

68. *Mode of estimating its Amount.*—The temperature of celestial space must be lower than that of the polar regions during the
coldest months of the year, for during winter these regions do
not lose all the heat received from the sun during the preceding
summer, and by means of winds there is a constant interchange
of heat between the polar and equatorial regions of the earth.
Now at Jakutsk, in Siberia, the mean temperature of the month
of January is 44° below zero. Moreover, from October to November, the temperature of that place sinks 34°; from November to December it sinks 18°, and from December to January 6°.
If the sun's heat were to be permanently withdrawn, the temperature would doubtless fall still lower than is now observed in
January, probably as low as −60°. We can not then suppose
the temperature of space to be higher than −60°.
Many different methods have been employed for estimating the
temperature of space. The average of the estimates of several
distinguished philosophers makes it as low as −80°.

69. *The Atmosphere a regulator of Temperature.*—The atmosphere serves as a regulator of the sun's heat. During the day it
absorbs a portion of the sun's rays, by which it is warmed, and as
it expands a part of the heat becomes latent. During the night
the air intercepts a part of the rays emitted by the earth, and as
it cools it contracts, and restores to the sensible condition the la-

tent heat which it had absorbed during the day. Without an atmosphere we should experience during the day an excessive heat from the sun's rays, no portion of which would be intercepted, and during the night an intense cold resulting from the unobstructed radiation of heat into space.

TEMPERATURE OF THE EARTH AT DIFFERENT DEPTHS.

70. *Means of Observation.*—For the purpose of measuring the variations of temperature beneath the surface of the earth, thermometers with very long stems have been buried at different depths in the ground, the stem being of such a length as to rise above the surface of the earth, so that the temperature can be observed without disturbing the position of the thermometer. For convenience of comparison, it has generally been agreed to adopt a uniform system, and thermometers have been buried at depths of 24, 12, 6, and 3 French feet.   [A French foot is about $\frac{1}{18}$th greater than an English foot.]   From twenty to thirty years ago, thermometers were buried at these four depths at Brussels, Edinburg, Greenwich, and other places; and other thermometers were also buried at depths less than three feet.   At first, these thermometers were observed several times each day, but afterward once a day or once a week.

71. *Range of the fluctuations of Temperature.*—Since the earth is a bad conductor of heat, the range of the fluctuations of temperature rapidly diminishes as we descend below the surface.   At a certain depth the diurnal variations of temperature disappear, and at a greater depth the annual variations also disappear. These depths vary as the square root of the period compared. The annual variations disappear at a depth 19 times greater than the diurnal variations; 19 being nearly the square root of 365, the number of days in a year.   In Europe generally, the diurnal variations are not sensible to a greater depth than $3\frac{1}{2}$ feet; but the depth varies somewhat with the latitude and the conducting power of the soil.

At the depth of three feet the annual range of temperature is less than half what it is at the surface; at the depth of twelve feet it is less than one fourth, and at the depth of twenty-four feet it is less than one tenth what it is at the surface.

72. *Stratum of Invariable Temperature.*—At a certain depth the annual variations of temperature become insensible; that is, we find a temperature which is invariable from summer to winter. This depth depends upon the extreme range of the temperature of the air. In Europe it is from 80 to 100 feet beneath the surface. A thermometer which has been kept for 75 years in the vaults of the Observatory at Paris, at the depth of 91 feet below the surface, has not varied more than half a degree during the entire interval.

The annual mean of the temperatures observed at different depths is very nearly the same as that of the air. Hence we are furnished with a convenient means of determining nearly the mean temperature of any locality, and this method is one of great value to scientific travelers.

73. *Time of Maximum and Minimum Temperature.*—Since the earth is a bad conductor of heat, the heat of the sun penetrates the ground slowly, and the highest temperature of the year occurs later and later the deeper we descend below the surface. At the depth of twelve feet the maximum temperature of the year does not occur until October, and the minimum occurs in April. At the depth of twenty-four feet, the maximum occurs in December, and the minimum in June or July. These dates vary somewhat in different countries, being dependent upon the conducting power of the soil.

The maximum of daily temperature also occurs later the deeper we descend, requiring three hours to penetrate to a depth of four inches.

74. *Increase of Temperature with the Depth.*—Below the depth of 100 feet from the surface, we find an invariable temperature throughout the year; but this temperature is not the same as the mean temperature at the surface. Numerous observations have been made in different parts of the globe, and they invariably indicate that the mean temperature increases with the depth. These observations have been extended to very great depths by means of mines and artesian wells. An artesian well consists of a shaft of a few inches in diameter, bored into the earth till a spring is found. To prevent the water from being carried off by the adjacent strata, a tube is generally inserted, exactly fitting the bore from top to bottom, and through this tube the water rises to the surface. Artesian borings have been made in Europe to a depth of more than 2300 feet below the surface, and some of the mines are more than 2000 feet deep.

In Europe the average increase of temperature deduced from mines and artesian wells is one degree for a descent of 52 feet.

75. *Rate of Increase in the United States.*—Some very deep borings have been made in the United States. An artesian well in Charleston, South Carolina, has a depth of 1000 feet; one in Louisville, Kentucky, has a depth of 2086 feet; a third in St. Louis, has a depth of 2200 feet; and a fourth in Columbus, Ohio, has a depth of 2575 feet. The boring at Louisville indicates an increase of temperature of one degree for every 76 feet; and that at Columbus gives an increase of one degree for every 71 feet. The mean of these two experiments gives an increase of one degree for every 73 feet, which is less than the rate of increase in Europe.

76. *Stratum of Frozen Earth.*—Throughout nearly the whole of the Arctic circle the mean temperature is considerably below 32°, and this is also the mean temperature of the surface of the earth. Now, in the polar regions, the earth in summer only thaws to a depth of three or four feet. Below this line is a stratum of permanent frost, whose depth increases as we advance northward, the lower limit being determined by the increase of temperature explained in Art. 74. At Jakutsk, latitude 62° 2′, it has been determined by actual excavation that the earth is frozen to a depth of 382 feet. In the polar regions, therefore, wells are impossible, unless they are sunk to a depth below that of the permanent frost.

77. *Temperature of the Earth at great Depths.*—If the temperature of the earth at great depths increases at the same rate as near the surface, at a depth of two miles the temperature must exceed that of boiling water, and at a depth of less than a hundred miles the rocks must be in a state of fusion. We are thus led to the conclusion that, with the exception of a comparatively thin crust upon the surface, the entire mass of the earth is probably in a state of igneous fusion.

78. *Information furnished by Volcanoes.*—This conclusion is confirmed by the phenomena of volcanoes. At numerous points upon the earth's surface we find volcanoes which frequently eject immense masses of melted rock, and which, without doubt, at all times contain large quantities of rock in a state of fusion. Volcanoes, extinct or active, border the Pacific Ocean from Cape Horn to the Arctic circle; thence they extend in a line to Asia, and along the coast of Japan to the Philippine Islands, New Guinea, and New Zealand; and they constitute half of the islands of the Pacific Ocean. Volcanoes occur also in Central and Western Asia; in Southern, Central, and Southwestern Europe; in Iceland and the West Indies. Volcanoes therefore are so numerous (their number exceeding 500) as to indicate that a considerable portion of the interior of the earth must be in a state of fusion.

Some have doubted whether the whole interior of the earth is in a state of fusion, and are disposed to admit only the existence of interior seas of liquid rock.

79. *Observations of Hot Springs.*—At many places remote from any active volcano we find natural springs which emit water of a very high temperature. Many of the springs of Germany have a temperature of 140 to 150 degrees, and one has a temperature of 167°. At New Lebanon, N. Y., is a spring whose temperature is 25° above the mean temperature of the place. A spring in Virginia has a temperature of 102°, another in North Carolina has a temperature of 125°, while one in Arkansas has a temperature of 148°. Near San Francisco is a spring which perpetually emits boiling water, and there is a similar one near the eastern boundary of California. These springs probably rise from great depths, and are proofs of the increasing temperature of the earth as we descend below the surface.

80. *Temperature of Ordinary Springs.* — The ordinary springs and wells of a country afford a convenient means of determining approximately its mean temperature. The mean temperature of the water proceeding from springs is nearly that of the strata from which they rise. Hence the water from deep springs preserves throughout the year a nearly uniform temperature, and this is generally a little *above* the mean temperature of the air. This

difference may amount to five or six degrees; and, on the contrary, the mean temperature of springs is sometimes a little *below*
that of the air. The temperature of springs is modified by the
temperature of the rain which supplies them. In those places
where the rain falls chiefly in summer, the mean temperature of
springs should be *higher* than that of the air, but it should be
lower in those countries where the rain falls chiefly in winter.
Hence great caution is required·in deducing the mean temperature of a place from the temperature of its springs.

81. *Low Temperature of certain Wells.*—In some wells the mean
temperature of the water is considerably below the mean temperature of the place. In ordinary wells the water is in continual
circulation, the water of the well flowing off by underground
streams, while fresh water flows in through similar channels.
Thus throughout the year the water of the well preserves nearly
the temperature of the earth at the same depth; and a few observations of such a well will furnish very nearly the mean temperature of the place. But in some wells there is very little circulation, the same water remaining in the well for a long time
with but trifling change. Now, since cold air is heavier than
warm air, the cold air of winter descends into the well, and communicates its own temperature to the water in the well. The
water thus becomes chilled, and it may even freeze, as actually
happens to many wells of New York and New England. When
considerable ice once forms in a well, it must remain for a long
time unmelted, because in summer the warm external air can not
displace the heavier cold air of the well. Under such circumstances, ice has been known to continue till after midsummer; and
the mean temperature of such a well may be several degrees below the mean temperature of the place.

82. *Remarkable Examples.*—In Brandon, Vt., is a well 34 feet
deep, in which, during the winter, ice forms six or eight inches
in thickness, and does not entirely disappear until the close of the
succeeding summer.

In Owego, N. Y., was formerly a well 77 feet in depth, where
ice formed during the winter, and has been known to continue
until near the close of July.

83. *Natural Ice-houses.*—In hilly countries we sometimes find secluded spots where the ice which accumulates in winter is so protected against the action of the sun in summer that it remains unmelted till August, or perhaps even through the year. The springs which flow from such places may show a temperature but little above 32°, even in midsummer. Several examples of this kind are found in New England, and still more remarkable examples are found in the mountainous districts of Europe. On the western bank of Lake Champlain, near the village of Port Henry, is an iron mine in which the ice accumulates in winter, and does not entirely disappear during the subsequent season. In Meriden, Conn., is a rocky ledge of little elevation, where the ice of winter remains unmelted until the succeeding August.

In the eastern part of France (Besançon), at an elevation of less than 3000 feet above the sea, is a cavern where the ice has been known to lie unmelted for more than a century.

84. *Temperature of the Sea.* — To determine the temperature of the sea at different depths we require some kind of self-registering thermometer. The instrument employed for such observations in the U. S. Coast Survey is Saxton's metallic thermometer.

This instrument consists of a compound coil or helix about six inches in length, formed of two stout ribbons of silver and platinum, with an intermediate thin plate of gold, all soldered together, the silver being on the inside of the coil. One end of this coil is firmly attached to the base of a cylinder, while the other end is fastened to a brass stem passing through the axis of the coil. When the temperature rises, the curvature of each spiral diminishes, because the silver expands more than the platinum; and when the temperature declines, the curvature of each spiral increases. The coil therefore winds and unwinds with the variations of temperature, and this motion gives rotation to the brass stem. This motion is registered upon the dial of the instrument by an index which pushes before it a registering hand, moving with friction barely sufficient to retain its place when thrust forward by the index of the thermometer. The instrument may thus be made to register both the highest and lowest temperatures to which it has been exposed.

D

85. *Temperature at the Surface of the Sea.*—The surface of the sea becomes heated less readily than the earth: 1st, because the rays of the sun penetrate the ocean to a considerable depth, and therefore produce less effect at the surface; 2d, because water has a much greater capacity for heat than dry earth; and, 3d, because, by the agitation of the sea, there is a perpetual mingling of the surface water with the lower strata. The surface also becomes cooled very slowly for the same reasons, and also because, when the particles of the surface are cooled, they descend, to be replaced by warmer particles from beneath.

Hence the diurnal variations of the temperature of the sea are quite small, amounting to only two or three degrees in the torrid zone, and 4° or 5° in the temperate zones. The minimum occurs about sunrise, and the maximum about noon.

Near the middle of the Atlantic Ocean, under the equator, the mean temperature of the sea is 80°.4. As we recede from the equator, the temperature of the sea declines somewhat less rapidly than the land, the mean temperature of the middle of the Atlantic being about four degrees warmer than the western coast of Africa and Europe.

The entire range of temperature for the middle of the Atlantic during the year, near the equator, is about 10°; near latitude 30° it is 15°; near latitude 40° it is 20°, and near latitude 50° it is 24°, which is scarcely one half the annual range of temperature of the most equable climates in the same latitude on land.

86. *Temperature at different Depths.*—Between the tropics the temperature of the sea decreases as we descend, at first rapidly, but afterward more slowly, to the depth of over 1000 fathoms, where the thermometer has been found to indicate 36°. Beyond latitude 25°, the decrease of temperature with the depth is less rapid; and beyond latitude 65°, during winter, the temperature sometimes increases as we descend. When the temperature of the surface-water was 28°, the temperature at the depth of 700 fathoms has been found to be 36°.

In very deep water, all over the globe, there is found to prevail a uniform temperature of 36° to 39°. The depth at which this temperature is found is about 7200 feet at the equator, and about 4500 feet in the highest accessible latitudes.

87. *Currents of the Sea.*—On the surface of the Atlantic Ocean, near the equator, there is a current setting westward, which divides where it meets the projecting coast of South America, one portion turning northward and the other southward. The former gives rise to the Gulf Stream, which travels along the coast of the United States to latitude 45°, whence a portion proceeds northeastwardly between Iceland and the British Islands, and the other portion descends along the western coast of Europe and Africa, and rejoins the equatorial waters.

The Brazil current coasts along the South American shore, and in the South Atlantic makes a circuit somewhat similar to that of the Gulf Stream in the north.

In the Pacific Ocean, a current setting westward prevails throughout the whole of the equatorial belt until near the Asiatic coast, where, as in the Atlantic, it divides, and one portion, called the Japan current, imitates in the North Pacific the course of the Gulf Stream in the North Atlantic. The larger portion of the equatorial current is, however, carried southward to sweep the northern and western coast of Australia.

At the bottom of the ocean there prevail *counter-currents*, which carry from the poles toward the equator the cold waters of the Arctic Seas. The existence of these currents is perceived whenever we sink a weight to a great depth by means of a long cord. This is the cause of the low temperature prevailing in tropical regions near the bottom of the ocean.

88. *Temperature of Banks.*—Where the sea is shallow the water is generally found somewhat colder than in the adjacent open ocean, the difference frequently amounting to ten degrees or more. This change of temperature is very noticeable over the Banks of Newfoundland, in contrast with the Gulf Stream, which flows near their eastern margin, where we frequently find a change of temperature of 33° within a distance of 300 miles. Thus a thermometer may frequently give warning of approaching land in the darkness of night, when nothing else would indicate it.

This low temperature over banks has been ascribed to the *under-current* from the polar regions toward the equator, which in deep water is only found at great depths, but which in shallow water is partially forced upward, so as to affect somewhat the temperature at the surface.

89. *Polar Ice.* — From latitude 40° to 50°, during winter, the water of the ocean freezes somewhat near the shore; but it is only in the polar regions that we find firm ice at a great distance from the land. Sea water freezes at a temperature of $27\frac{1}{2}°$, and since, during winter, the mean temperature of the polar regions is considerably below zero of Fahrenheit, ice forms even in the open sea with great rapidity, and sometimes attains a thickness of twenty-five feet. In the spring of the year this ice is partially dissolved; it is then broken up by tides and currents, and by northerly winds is driven into the open sea, sometimes forming a field of ice 100 miles in length and 50 miles in breadth, with a thickness of 20 or 25 feet. During the months of May and June this ice is annually encountered in immense fields off the coast of Newfoundland, near the track of vessels from New York to Liverpool.

In connection with these immense fields of comparatively thin ice are generally found some masses of ice called *icebergs*, sometimes rising 200 feet above the water, and descending to a depth of 1000 feet beneath the surface. These masses are detached from the coasts, around which, in winter, the ice accumulates in cliffs of vast height and extent. The largest of them are detached portions of vast glaciers, such as abound on the precipitous coast of Greenland and Spitzbergen, which were broken off either by their own weight or the action of the waves, and then transported by winds and currents to the lower latitudes. Fig.

Fig. 19.

19 represents an iceberg encountered some years since near the Cape of Good Hope.

90. *Temperature of Lakes and Rivers.*—The temperature of lakes exhibits changes much greater than those of the ocean. The surface may freeze in winter, while in summer the temperature may rise to 77°. In deep lakes, at a certain depth, we find a constant temperature of about 39°, this being the temperature of water at its maximum density. Since the warm water of the surface descends as fast as it becomes cooled, the surface of a lake can not freeze until the entire mass has fallen to the temperature of 39°, unless under the influence of very sudden and severe cold.

In rivers the constant agitation of the water tends to render the temperature uniform throughout. Hence the temperature at the surface would not change very greatly during the year were it not for the diminished flow in summer, which leaves but a thin stratum of water to be acted upon by the sun. During winter congelation can not take place until the entire mass is cooled to 32°, with the exception perhaps of deep cavities.

91. *Anchor Ice.*—Ice sometimes forms upon stones and other objects at the bottom of rivers when the surface water is not frozen, and this is called anchor ice. Such ice may form under the following circumstances. During a period of severe cold the water of a river may sink below 32° from top to bottom throughout, and the surface water not congeal, because it is kept in constant agitation, while the water at the bottom, being more quiet, may congeal. The ice thus formed at the bottom forms a nucleus about which the congelation continues and extends. When the ice becomes quite thick, its buoyant force may overcome its adhesion to objects at the bottom, and it rises to the surface. A slight elevation of temperature, causing a partial fusion, may also detach it from the bottom.

Anchor ice never forms at the bottom of tranquil water, because congelation commences at the surface, while the temperature of the bottom is above 32°.

CHAPTER III.

THE MOISTURE OF THE AIR.

92. *How Water is converted into Vapor.*—If during summer we expose to the sun's rays a vessel containing water, we find that the water rapidly diminishes, and in a few days entirely disappears. The water seems to have been annihilated, but in fact it has been converted into vapor, which is diffused through the atmosphere. This vapor is entirely invisible, but by the application of cold we may condense it, and reduce it again to the form of water. Thus in summer, if we pour cold water into a metallic vessel, we find that the outside of the vessel, which was previously quite dry, soon becomes covered with moisture. This moisture does not come from the inside of the vessel. It is simply the vapor of the air, condensed by coming in contact with a cold surface.

The vapor of the air may be condensed in a similar manner at all seasons of the year. The phenomenon is most frequently noticed in summer, because then the temperature of the air rises highest above that of the water which we are accustomed to use. But at any period of the year, if the water be not already cold enough, by adding to it ice, and if necessary salt, we may condense the moisture of the air even in the coldest weather.

93. *How Vapor is sustained in the Air.*—The atmosphere always contains vapor of water. This vapor is not sustained in the air like water in a sponge, nor does it float in the air like small particles of dust, but it penetrates between the particles of the permanent gases which compose the atmosphere, and sustains itself precisely in the same manner as they do. If we exhaust all the air from a close vessel, and introduce into it a quantity of water, a portion of the water will immediately pass into the state of vapor, which will fill the entire vessel. Indeed, with the exception of the facility with which it is reduced to the liquid state, vapor of water has precisely the same properties as oxygen or nitrogen.

If into a close vessel containing atmospheric air perfectly dry we introduce a quantity of water, vapor will be formed of the same tension as if the vessel were previously void. The only difference will be that in a vacuum the maximum tension of the vapor will be attained instantly, while in a vessel filled with gas a certain time will be required to produce the same result.

94. *Amount of Evaporation Measured.*—The amount of evaporation from the earth's surface is measured by placing a vessel of water in the open air, and determining the loss of water from day to day. The vessel usually employed for this purpose is a cylinder from six to twelve inches in diameter. It is nearly filled with water, the quantity having been previously weighed or measured; it is then placed out of doors, freely exposed to the action of the atmosphere. At the end of twelve or twenty-four hours the water is again measured, and the loss of water shows the amount of evaporation that has taken place. If rain has fallen between the two observations, the amount collected in the rain gauge must be deducted from the quantity in the evaporating gauge. The wire cage around the gauge, Fig. 20, is to prevent animals, birds, etc., from drinking the water.

Fig. 20.

From observations continued for nine years at London, Howard determined that the average amount of evaporation was thirty inches annually, although the annual fall of rain at that place is only twenty-five inches.

95. *Rate of Evaporation Variable.*—The rate of evaporation depends greatly upon the exposure of the evaporating dish. If the vessel be freely exposed to the sun and wind, the amount of evaporation will be greater than that which takes place from the surface of the earth; but if the vessel be very much sheltered, the result will be too small. The total evaporation from the earth's surface in a year must be equal to the total precipitation in the form of rain, snow, dew, etc.; but hitherto the relative amount of evaporation from the ocean and from the land has not been accurately determined.

Evaporation is accelerated by a brisk wind. The vapor which rises from water and pervades the surrounding air, is carried off by a wind which brings a fresh body of air in contact with the water.

96. *Evaporation at all Temperatures.*—Evaporation proceeds at all temperatures, even the lowest. If during the coldest weather of winter we weigh a lump of ice, and then expose it in the open air on a clear day upon the north side of a building, we soon find that the ice has lost weight. So also in winter a large mass of snow often disappears without any appearance of liquefaction. Evaporation proceeds, although at a diminished rate, even when the thermometer stands below zero of Fahrenheit.

HYGROMETERS.

97. Any instrument adapted to measure the amount of moisture in the air is called a *hygrometer.* An instrument which simply indicates changes of humidity is called a *hygroscope.* All organic substances are affected by moisture, which generally increases their dimensions. Thus porous wood expands with an increase of moisture, and contracts when deprived of moisture. A strip of such wood may be employed as a hygroscope, but it is not sufficiently sensitive for any useful purpose. A thin shaving of whalebone, or a single hair, is much more sensitive. A hair will vary to the amount of one fiftieth of its entire length by simple change of moisture.

Fig. 21.

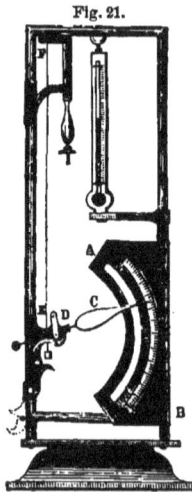

98. *Saussure's Hygrometer.*—This instrument consists of a metallic frame, to the top of which is attached one extremity of a hair, E F, whose lower extremity is wound around a small wheel. To the axis of this wheel is attached an index, C, whose extremity traverses a graduated arc. When the moisture of the air increases, the hair lengthens and the index descends; when the moisture decreases, the index rises.

To graduate the instrument, we determine two fixed points, viz., that of *saturation* and that of *extreme dryness.* To obtain the first point, we place the instrument under a close vessel containing water, and mark the position of the index. For the point of absolute dryness, we place the instrument in a dry vessel containing quick-lime. The interval between these fixed points is divided into one hundred parts, which are called the degrees of the hygrometer.

In Babinet's hygrometer, the variations in the length of the hair from day to day are measured by means of a microscope attached to the frame of the instrument.

The hair hygrometer is a very imperfect instrument. It is essential to a perfect hygrometer that two instruments made independently in distant countries should agree with each other. But it is found that two instruments made with different hairs, or with hairs differently prepared, may differ in their indications five degrees. Even the same hygrometer undergoes a gradual change, since the length of the hair increases from the continued pressure of the weight which it supports. This instrument is therefore so unsatisfactory that it has been entirely discarded in scientific researches.

99. *Dew-point Defined.*—The amount of vapor in the air may be measured with great accuracy by noting the temperature at which moisture begins to be condensed on a cold vessel. The moisture thus deposited is called *dew,* and the temperature at which this deposition begins is called the *dew-point.* The dew-point, then, may always be determined by cooling a metallic vessel until dew begins to appear upon its surface, and noting by a thermometer the temperature of the vessel. This experiment, however, requires considerable time, and various contrivances have been proposed to facilitate it.

Fig. 22.

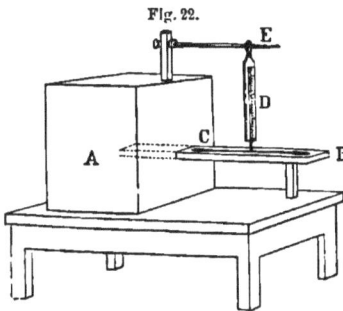

100. *Bache's Hygrometer.* — When it is required to determine the dew-point frequently at short intervals, the following apparatus, invented by Professor Bache, is very convenient. A small metallic box, A, is filled with a mixture of salt and snow, by which means its temperature is reduced to about zero. From the side of the box projects a polished metallic bar, B, having on its upper side a groove, C, containing mercury, in which is immersed the bulb of a thermometer, D, which is suspended from a support, E, so that the thermometer is movable along the groove. One end of the bar, B, has a very low temperature, while the other is but little below that of the surrounding air. That portion of the bar whose temperature is below the dew-point will be covered with moisture, while the other part will be dry, and the two portions will be separated by a well-defined bounding line. By placing the bulb of the thermometer, D, opposite to this line, we may immediately determine the temperature of the dew-point. When only an occasional observation of the dew-point is desired, this instrument is inconvenient, because it requires considerable time to prepare it for experiment.

101. *Daniell's Hygrometer.*—This instrument is more convenient than Bache's when only an occasional observation is to be made.

Fig. 23.

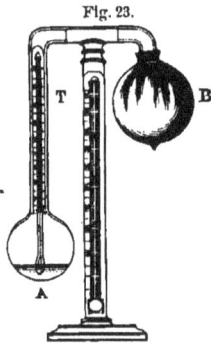

It consists of two glass bulbs, A and B, about three fourths of an inch in diameter, connected by a small tube, which is bent in two places at right angles, and the whole is hermetically sealed. The lower bulb, A, which is made of dark-colored glass, is about half filled with ether, and contains a small thermometer, T. The upper bulb, B, is covered with a piece of fine muslin. If we pour ether upon the ball B, the ether will rapidly evaporate and produce cold, condensing the vapor of ether which previously filled the ball B. The ether in the ball A, being relieved from the pressure of the vapor upon it, now rapidly evaporates, and its temperature falls, as is shown by the depression of the thermometer, T. If this depression be sufficient, the vapor of the atmosphere will be condensed on the outside of the ball, and the state of the thermometer, T, at that instant will indicate the dew-point.

This instrument is ordinarily very convenient for use, but when the atmosphere is very dry it requires ether of the best quality, and some dexterity in manipulation, to obtain a deposit of dew.

Fig. 24.

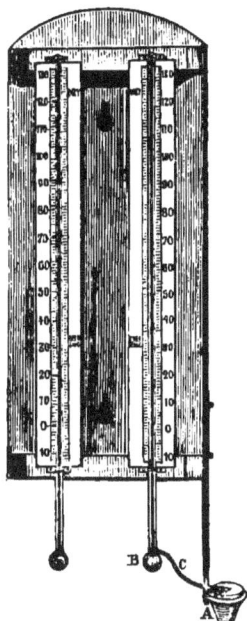

102. *Wet-bulb Thermometer.*—The hygrometer which, on account of its convenience, is now most generally used, is the wet-bulb thermometer. It consists of a common thermometer, with its bulb, B, Fig. 24, covered with a piece of thin muslin, and kept constantly moistened with water by means of loose cotton threads communicating with a cup of water, A. The evaporation of the water produces cold, and this thermometer habitually stands lower than a dry thermometer similarly exposed. This depression strictly measures only the *evaporating* power of the air; yet, as the latter depends upon the amount of moisture present in the air, the depression of the wet-bulb thermometer indirectly measures the humidity of the air.

103. *Dew-point deduced from the Wet Bulb.*—The difference between the temperature of the air and that of the dew-point is called the *complement* of the dew-point. When the air is saturated with moisture the complement of the dew-point is zero.

From the comparison of a great number of observations with Daniell's hygrometer, combined with simultaneous observations of the dry and wet bulb thermometers, a method has been discovered by which the dew-point may be deduced from the readings of the wet-bulb thermometer. The ratio of the complement of the dew-point to the depression of the wet-bulb thermometer is a variable one. When the temperature of the air is 53°, the difference between the readings of the dry and wet bulb thermometers is one half the complement of the dew-point; at 33° it is one third; at 26° it is one sixth; and at lower temperatures the ratio is still less. Table XXV., p. 273, furnishes the factors by which the dew-point may be deduced from the readings of the wet-bulb thermometer for any temperature of the external air.

104. *Weight of Vapor determined.*—The elastic force of the vapor present in the air, that is, the pressure which it exerts, is indicated by the dew-point. Dalton constructed a table showing for every degree of temperature the corresponding elastic force of vapor, and this table has since been brought to great perfection.

When the dew-point is at
$\left\{ \begin{array}{l} 32° \\ 40° \\ 50° \\ 60° \\ 70° \\ 80° \end{array} \right.$
the pressure of the vapor in the air will sustain a column of mercury
$\left\{ \begin{array}{l} 0.181 \text{ inch in height.} \\ .248 \quad \text{``} \quad \text{``} \\ .361 \quad \text{``} \quad \text{``} \\ .518 \quad \text{``} \quad \text{``} \\ .733 \quad \text{``} \quad \text{``} \\ 1.023 \quad \text{``} \quad \text{``} \end{array} \right.$

A more extensive table is given on page 276. With the assistance of such a table, from the indications of either of the hygrometers already described, we can deduce the elastic force of the vapor present in the air, and hence we may determine its weight.

105. *How the Humidity of the Air is denoted.*—The character of a climate, whether it is to be regarded as dry or humid, does not depend simply upon the absolute amount of vapor present in the air. Its humidity is expressed by the ratio which the amount of vapor actually present in the air bears to the amount which the air would contain if it was saturated. Thus, suppose the temperature of the air to be 60°, while the dew-point is 50°. The pressure of the vapor in the air according to the table in Art. 104, is ·36 inch; but if the atmosphere were saturated with moisture, that is, if the dew-point were 60°, the pressure of the vapor would be .52 inch. Hence the air contains 70 per cent. of the amount of vapor which it would contain if it were saturated, and its humidity may be represented by the number 70. See Table XXVI.

In this manner we find that at Philadelphia the average humidity of the air is 73; that is, the air, on an average, contains about three fourths of the vapor required for its saturation. At St. Helena, the mean humidity of the air is 88, while at Madrid it is only 62. Near great bodies of water the atmosphere generally contains more moisture than it does over the interior of continents.

106. *Extremes of Humidity.*—In different localities and at different times we meet with every variety of condition, from perfect humidity to almost absolute dryness. In ordinary pleasant weather, the complement of the dew-point is from 10° to 15°. Occasionally, at Philadelphia, it amounts to 25° or 30°, and it has been observed as high as 45°. · In India, the temperature of the air has been known to rise 61° above the dew-point; and it is said that in California the temperature has been observed 78° above the dew-point, in which case the atmosphere contained only six per cent. of the vapor required for its saturation.

107. *Diurnal Variation in amount of Vapor.*—The amount of vapor present in the air is subject to great fluctuations, some of which are periodical. One of these fluctuations has a period of one day. At Philadelphia, the amount of vapor present in the air is least about an hour before sunrise, from which time the amount increases uninterruptedly until a little before sunset,

Fig. 25.

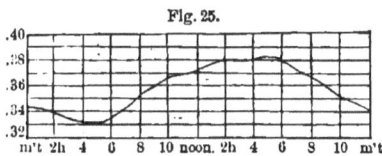

after which it decreases uninterruptedly until the next morning. The mean diurnal variation amounts to one eighth part of the average amount of vapor. Fig. 25 shows the diurnal variation at Philadelphia, the numbers on the left indicating, in inches of mercury, the pressure of the vapor at the hours given at the bottom of the figure.

The cause of this variation is obvious. As the sun rises and the heat of the day increases, more water is evaporated from the· ocean and the moist earth, and the amount of vapor in the air increases. During the night a portion of this vapor is condensed in the form of dew and hoar-frost; that is, the amount of vapor present in the air is least a short time before sunrise, and greatest a short time before sunset.

108. *Annual Variation in amount of Vapor.*—There is an annual variation in the amount of vapor present in the air. At Philadelphia the vapor present in the air is least in January and greatest in July; the amount in July being more than four times as great as in January. This is evidently the effect of the sun's heat producing a more rapid evaporation in summer than in winter.

109. *Influence of Elevation.*—The humidity of the air generally diminishes as we rise above the surface of the earth. From a large number of balloon ascensions near London, it has been found that when the sky is clear there is a slight increase of humidity until we reach an elevation of 3000 feet, and afterward a gradual decrease to 23,000 feet, where the humidity is expressed by 16. When the sky is overcast the increase of humidity up to the height of 3000 feet is very slight, after which there is generally a decrease, but very irregularly up to 23,000 feet. At the highest elevations at which observations have been made, the air has never been found entirely free from vapor of water.

110. *Diurnal Variation of the Barometer explained.*—We have seen, Art. 22, that the height of the barometer is subject to a diurnal fluctuation. This fluctuation is a complex effect, depending partly upon a change in the amount of vapor, and partly upon a change in the weight of the gaseous atmosphere. It is only when we separate these two effects that their cause can be clearly understood. We have seen that there is a diurnal variation in the amount of vapor present in the air, and that this variation depends upon the heat of the sun. If from the entire height of the barometric column we subtract the pressure of the vapor, the remainder will represent the pressure of the gaseous portion of the atmosphere.

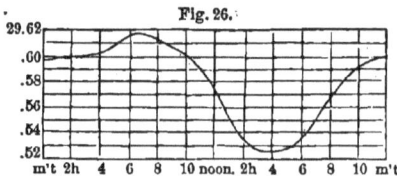

Fig. 26.

111. *Diurnal Variation of Pressure of the Gaseous Atmosphere.*—At Philadelphia the pressure of the gaseous atmosphere is greatest about an hour after sunrise, from which time the pressure diminishes uninterruptedly until about 4 P.M., after which the pressure increases uninterruptedly until the next morning, as shown by the curve in Fig. 26.

This fluctuation is evidently the effect of the sun's heat. As the heat of the day increases, the atmosphere becomes warmed, it expands in volume, and swells up to a height greater than it had during the night. The upper portion therefore flows off laterally in all directions to places where the height of the atmosphere is less; by which means the pressure of the air is diminished, and

. . the barometer falls. During the night the temperature declines, the air contracts in volume, its height sinks below that which existed during the day, and the defect is supplied by air which flows in from regions where a higher temperature prevails. The pressure of the air is thereby increased, and the barometer again rises.

112. *Why the Barometer shows two daily Maxima.*—The pressure of the vapor and that of the gaseous atmosphere have each but one daily maximum and minimum. But the motions of the vapor and of the gaseous atmosphere following different laws, and their maxima occurring at nearly opposite hours of the day, the *sum* of their effects, or the total pressure as shown by the barometer, exhibits two daily maxima and minima, which occur at different hours from the maximum and minimum of temperature.

113. *Annual Variation of Pressure of the Gaseous Atmosphere.*— At Philadelphia the pressure of the gaseous atmosphere is greatest in January, from which time it diminishes uninterruptedly until July, after which it increases uninterruptedly until the succeeding January. A similar remark is applicable to nearly every part of the globe, with this exception, that the difference between the winter and summer pressures is very unequal in different countries. At Philadelphia and Boston this difference amounts to half an inch, but throughout nearly the whole of Central Asia the difference amounts to an entire inch and upward, while under the equator it is scarcely appreciable.

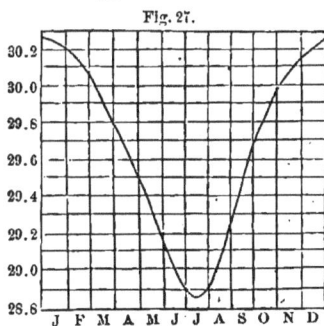

Fig. 27 shows the annual curve of pressure of the gaseous atmosphere at Pekin, in China.

This fluctuation in the weight of the gaseous atmosphere is due to the influence of the sun's heat, combined with the effect of the excessive rains on the mountain ranges of Central Asia. As the sun advances from the southern to the northern hemisphere,

the latter is heated and its atmosphere expands, while the former . .
is cooled and its atmosphere contracts. The atmosphere in the
northern hemisphere being thus rendered higher than in the
southern, the excess of air in the northern hemisphere flows over
to the southern ; in other words, the barometer is lowest in the
hemisphere where summer prevails, and highest in that where
winter prevails. The amount of this effect depends partly upon
the annual range of the thermometer. Over the great desert of
China the air in summer becomes unusually heated, the air above
it is expanded to a corresponding height, and flows off to the
colder portions of the southern hemisphere.

The remarkably low state of the barometer which prevails in
summer throughout a large part of Asia is probably due, in a
great degree, to the excessive rains on the mountain ranges of
Central Asia, in accordance with a general principle which will
be developed in Chapter VI.

In the temperate zones of Europe and America during summer, the increase in the amount of vapor is nearly equal to the
loss of weight sustained by the gaseous atmosphere, so that the
absolute height of the barometer remains nearly the same through
every month of the year.

CHAPTER IV.

THE MOTIONS OF THE ATMOSPHERE.

114. WIND is air in motion. The movements of the air are
proverbially variable and seemingly capricious, and it has been
supposed that they are not subject to any law. We shall find,
however, that the winds are subject to laws as definite as those
of the barometer or thermometer.

The direction of the wind is designated by the point of the horizon *from* which it blows. This direction is commonly indicated,
as in navigation, by the terms north, north by east, north-north-
east, etc. If we wish to indicate the direction with greater precision, we may employ degrees of azimuth, as in astronomy ; thus
a wind designated by N. 13° E. comes from a point 13 degrees to
the east of north. Sometimes it is found convenient to designate

the direction by degrees of the horizon reckoned continuously from 0° up to 360°.

For the purpose of investigating the laws which govern the movements of the atmosphere, we require some means of measur-. ing both the *direction* and *velocity* of the wind.

115. *How to determine the Direction of the Wind.*—Any instrument for measuring the direction of the wind near the earth's surface is called an *anemoscope*. The simplest anemoscope is the common vane. In order that the vane may give reliable results, particular care is required in its construction. A vane usually consists of a flat vertical plate, turning freely about an upright spindle. That part of the vane which is before the spindle, and is turned toward the wind, is called the head; the rest of the vane is called the tail. If a vane were made in the form of a rectangular plate of uniform thickness, and balanced upon its centre of gravity, the action of the wind upon the head would just neutralize its action upon the tail, and the vane would have no directive power. The directive power of the vane depends simply upon the *difference* of the wind's action upon the head and tail. The tail should therefore present a large amount of surface, and the head a small surface. Moreover, in order to maintain the spindle in an upright position with the least friction against its supports, the two ends of the vane should exactly balance each other.

The vane represented in Fig. 28 is designed to fulfill these conditions. It consists of a rod of iron, A B, three fourths of an inch in diameter, to one end of which is attached a pine board about

Fig. 28.

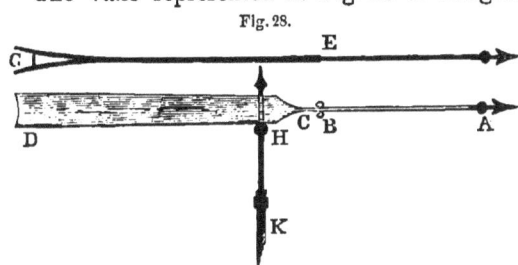

half an inch thick, one foot wide, and eleven feet long, and balanced by a sphere of iron or lead, A, attached to the other end of the rod. To give to the instrument more steadiness, the wooden part is made to consist of two boards inclined at a small angle, as shown in the section E G. The vane is attached to an upright

E

spindle, H K, which revolves freely, and the direction of the wind is measured by a graduated circle attached to the spindle.

116. *Self-registering Anemoscope.*—An anemoscope may be rendered self-registering in the following manner: Place a cylin-

Fig. 29.

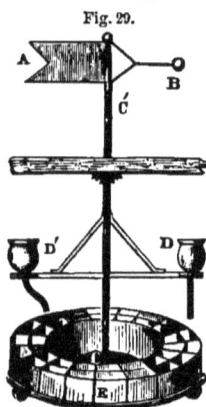

drical vessel beneath the revolving shaft C C′, which carries the vane A B, and let it be divided into a large number of equal compartments, as shown in Fig. 29. Attach to the shaft a funnel, D, filled with sand so arranged that in every position of the funnel the sand, as it flows out, shall fall into one of the compartments of the cylindrical vessel. The amount of sand which collects in the several compartments will indicate how long the vane is maintained in the corresponding positions. If there are eighteen compartments, each will correspond to an arc of twenty degrees.

A second series of compartments may be arranged in the same cylindrical vessel, and a second funnel, D′, be arranged like the first, for the purpose of balancing the weight of D.

117. *Woltmann's Anemometer.* — An instrument designed to *measure* the velocity or force of the wind is called an *anemometer*.

Woltmann's anemometer consists of a small wind-mill, to whose axis is attached an endless screw, which imparts motion to a toothed-wheel, while the number of revolutions is shown by an index. An observation consists in determining the number of revolutions made in one minute, when the sails are exposed to the action of the wind. In order to deduce the wind's velocity from such an observation, upon a calm day we travel with the apparatus on a carriage or a rail-car, and observe the number of revolutions made in going a known distance in a given time. The effect will be the same as if the instrument was at rest and the air in motion. In this manner we may construct a table showing the velocity of the wind corresponding to a given number of revolutions of the sails per minute.

118. *Whewell's Anemometer.*—Whewell's anemometer, Fig. 30,

Fig. 30.

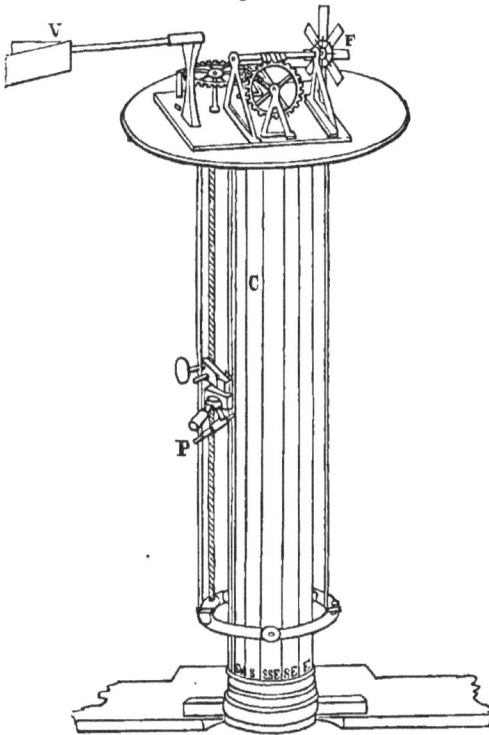

consists also of a small wind-mill, with complete apparatus for registering the total effect of the wind. The mill is mounted upon a vertical cylinder, C, about two feet high, and four inches in diameter, and around the cylinder is coiled a sheet of paper, ruled vertically, to indicate the points of the compass. The revolution of the arms of the windmill, F, gives motion to an endless screw, which causes a pencil, P, to descend along a vertical rod, and traces an undulating line upon the paper cylinder. When the pencil has reached the bottom of the paper (which ordinarily requires an interval of at least twenty-four hours), a new sheet of paper must be applied to the cylinder, and the pencil set back again at the top. The direction of the wind is indicated by the portion of the sheet upon which the pencil line is traced, and its velocity by the rate of motion of the pencil. Thus this instrument registers the amount of the wind's progress for every point of the compass.

119. *Robinson's Anemometer.*—Robinson's anemometer, Fig. 31, consists of four equal metallic cups, A, B, C, D, in the form of hemispheres, attached to two arms which cross each other at right angles, and are supported so as to turn freely about a vertical

Fig. 31.

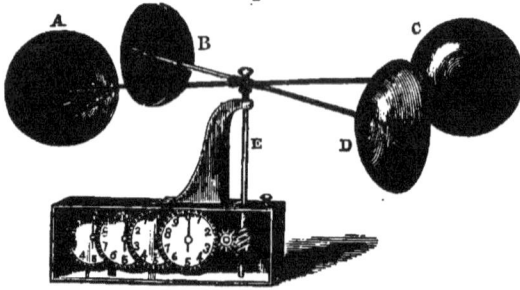

axis, E. The base of each hemispherical cup is in a vertical position; and since the action of the wind upon the concave side of one of these cups is greater than its action upon the convex side, a moderate breeze is sufficient to maintain the arms in continuous rotation. Dr. Robinson has proved that (making no allowance for friction) the centre of each hemisphere moves with one third the velocity of the wind, and thus this instrument measures directly the wind's velocity. The axis E carries an endless screw, which gives motion to a series of wheels which register the wind's progress up to 1000 miles.

Fig. 32.

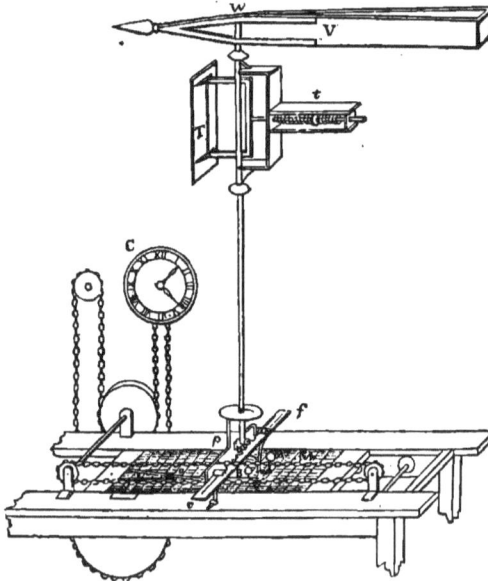

120. *Osler's Anemometer.*— Osler's anemometer, Fig. 32, registers both the direction and force of the wind. It consists of a large vane, V, supported upon a revolving spindle. Attached to the lower extremity of this spindle is a small pinion working in a rack, *ef,* which slides backward and forward as the wind turns

the vane, and to this rack is attached a pencil, A, which presses against a horizontal sheet of paper, ruled to indicate the points of the compass. This sheet of paper is moved forward uniformly by clock-work, C, at the rate of half an inch per hour, so that while the vane oscillates to and fro, the direction is registered on the sheet of paper, which also indicates the time at which each change took place. Fig. 33 shows the register made by the pencil in one day.

Fig. 33.

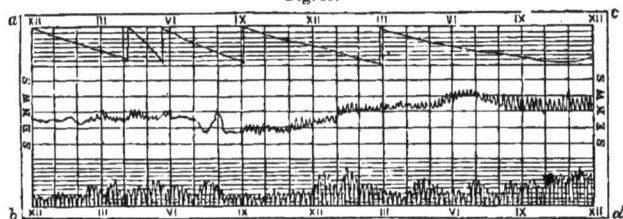

121. *How the Wind's Force is Measured.*—In order to measure the wind's force, a brass plate, T, two feet square is attached to the vane, so as always to be presented perpendicularly to the action of the wind. To the back of this plate is attached a spiral spring, which is compressed by the pressure of the wind against the plate, and the degree of compression of the spring affords a measure of the wind's force. By means of a connecting wire this square plate gives motion to a second pencil, B, which at each instant registers upon the same sheet the wind's force. At the end of twenty-four hours a new sheet must be applied, and thus each sheet indicates the direction and force of the wind for each instant during a period of twenty-four hours. The undulating line at bottom of Fig. 33 shows the register of the wind's force. The irregular line at the top of the same figure shows the amount of rain registered by an arrangement not here represented.

122. *How Velocity is deduced from Pressure.*—The indications of Osler's anemometer are expressed in pounds of pressure per square foot. In order to deduce from these results the velocity of the wind in miles per hour, we require a table showing the velocity of the wind corresponding to different pressures. The following table shows the velocity of the wind in miles per hour, corresponding to the pressure upon a square foot of surface, according to the experiments of Smeaton.

Velocity. Miles.	Pressure. Pounds.	Velocity. Miles.	Pressure. Pounds.	Velocity. Miles.	Pressure. Pounds.	Velocity. Miles.	Pressure. Pounds.
1	0.005	6	0.177	11	0.595	16	1.260
2	.020	7	.241	12	0.708	17	1.422
3	.044	8	.315	13	0.831	18	1.594
4	.079	9	.399	14	0.964	19	1.776
5	.123	10	.492	15	1.107	20	1.968

It will be seen that the wind's force varies as the square of its velocity. Thus, when the wind's velocity is 20 miles per hour, its force is four times as great as that of a wind blowing 10 miles per hour.

123. *Wind's Force represented by a Scale.*—When an observer has no anemometer, he should estimate the force of the wind as accurately as he is able, and it is recommended to indicate the wind's force by a series of numbers from 1 to 10, according to the following scale:

No.	Character.	Velocity in Miles per Hour.	Force in Pounds per square Foot.	No.	Character.	Velocity in Miles per Hour.	Force in Pounds per square Foot.
1	Just perceptible.	2	0.02	6	Very high wind.	45	10
2	Gently pleasant.	4	0.08	7	Strong gale.	60	18
3	Pleasant brisk.	12½	0.75	8	Violent gale.	70	24
4	Very brisk.	25	3.00	9	Hurricane.	80	31
5	High wind.	35	6.	10	Most violent hurricane.	100	49

The numbers in the preceding table have been deduced from a great variety of experiments. One mode of experimenting consists in noting the effects produced by a motion of the observer at different velocities during a clear day, as, for example, upon a railway train. Another method consists in measuring the velocity with which light objects, like a lock of cotton, are carried along by the wind.

124. *Average Velocity of the Wind.*—Observations with accurate anemometers have been made at several places in Europe, and at a few stations in America. It is found that at Philadelphia the mean velocity of the wind during the entire year is eleven miles per hour, being least in summer, when it is nine miles per hour, and greatest in winter, when it is fourteeen miles per hour. At Toronto, the average velocity of the wind is nine miles per hour.

At Plymouth, England, the average velocity of the wind is nine miles per hour; at Oxford and Greenwich it is ten miles per hour; and at Liverpool it is thirteen miles per hour. On the ocean, the mean velocity of the wind, as deduced from the average rate of sailing of ships, is estimated at eighteen miles per hour.

According to observations at Philadelphia, the mean velocity of the wind is least about sunrise. After sunrise the velocity rapidly increases, and becomes greatest at 2 P.M., after which it rapidly declines till 8 P.M., from which time it changes but little until sunrise, the pressure at noon being fully double that at midnight. Fig. 34 represents the average force of the wind at Philadelphia for each hour of the day, expressed in pounds pressure per square foot, as shown on the left of the figure.

Fig. 34.

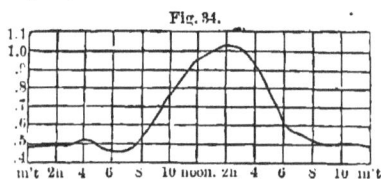

125. *Mean Direction of the Wind.*—Suppose a current of air coming from the north passes the point C, Fig. 35, with a velocity v, continued for a time t, the amount of air which passes will be measured by vt. If another current subsequently coming from the south moves with a velocity v' during a time t', the amount of air which passes will be measured by $v't'$, and the resulting motion will be the same as if a mass of air $vt - v't'$ passed the point C during the time $t + t'$. If then N and S represent masses of air coming from the north and south, the resulting motion will be represented by $N-S$. In like manner, if we consider two winds coming successively from the east and west, the resulting motion will be represented by $E-W$.

Fig. 35.

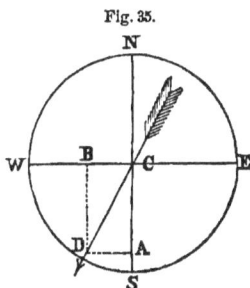

If we represent these results $N-S$ and $E-W$ by the lines C A, C B, we may easily determine their resultant, C D. The angle V which it makes with the meridian N S is given by the formula

$$\text{tang. } V = \frac{D\,A}{C\,A} = \frac{C\,B}{C\,A} = \frac{E-W}{N-S}.$$

A wind blowing from any intermediate point may be resolved into two others, one of which coincides with a meridian, and the other is perpendicular to it. A wind from the northeast may be resolved into two others, one in the direction of C S, represented by NE cos. 45°, and the other in the direction of C W, also equal to NE cos. 45°. A wind from the northwest, southeast, or south-west may be resolved in a similar manner. If then we consider .the winds from the eight principal points, and regard motion from N to S, or from E to W as positive, while we regard motion from S to N, or from W to E as negative, we shall have

$$\text{tang. } V = \frac{E - W + (NE + SE - NW - SW) \cos. 45°}{N - S + (NE + NW - SE - SW) \cos. 45°}.$$

The mean velocity of the resulting wind is given by the formula

$$C D = \frac{C B}{\sin. V} = \frac{E - W + (NE + SE - NW - SW) \cos. 45°}{\sin. V.}.$$

In most meteorological registers the velocity of the wind is not measured, or perhaps not even estimated, and we are obliged to assume that the average velocity of the wind is the same for all points of the compass, in which case N, S, E, etc., in the pre-ceding formula, represent simply the *number of times* that the wind has blown from each of these points.

The assumption that the winds from the different points of the compass blow with the same average velocity is not entirely cor-rect, and the error which may result from its adoption can only be determined by careful observations with an anemometer.

When the direction of the winds is given for more than eight points of the compass, we may resolve each wind separately into two rectangular components by means of a traverse table, in the same manner as we resolve a traverse in navigation. We then subtract the sum of all the southerly motions from the sum of all the northerly, and represent this difference by C A. We also subtract the sum of all the westerly motions from the sum of all the easterly motions, and represent this difference by C B. The resulting direction will then be given by the equation

$$\text{tang. } V = \frac{C B}{C A}.$$

126. *Wind's Progress represented by a Polygon.*—A geometrical figure to represent the total progress of .the wind for an entire year may be constructed as follows: Draw the line A B, Fig. 36, to represent a northwest direction, and, assuming any convenient

Fig. 36.

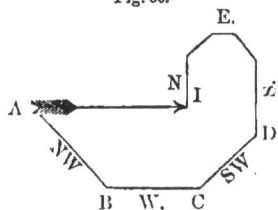

scale, make the length of A B to correspond to the northwest motion of the wind for the given time. Draw B C to represent a west direction, and make its length to correspond to the west motion of the wind upon the same scale as the preceding. In the same manner draw C D for the southwest wind, and so on for each of the other directions of the wind, and suppose the last line representing the north winds reaches to I. Then join A I, and this line will represent the direction and rate of the wind's total progress during the period embraced in the observations. The annexed figure represents the relative frequency of the different winds, according to observations made during twenty-five years, at about thirty academies in the State of New York.

By a series of observations with Osler's anemometer, it is found that at Philadelphia the actual progress of the wind is toward a point a little north of east, and at the average rate of about four miles per hour, or one hundred miles per day.

127. *Observations of the Wind's Direction.*—Although observations with accurate anemometers are not very numerous, yet observations of the common vane have been made to such an extent as to determine (if not the velocity of the wind) at least its average direction for nearly every part of the globe. In the northern hemisphere we have observations from about six hundred stations on land, at which the wind's direction has been recorded for periods varying from a few months to more than half a century, and amounting in the aggregate to nearly three thousand years of observations. We have also the log-books of ships which have penetrated nearly every sea, and which have been collected at the Observatory of Washington, furnishing more than three millions of observations, and embracing in the aggregate a period of more than three thousand years of observation. These materials are sufficient to indicate with considerable precision the average direction of the wind for every part of the northern hemisphere, whether over the continents or the ocean, at least as far as latitude 60°. Beyond latitude 60° observations are much less numerous; nevertheless, the observations which we have from this region are pretty uniform in their indications. In the southern hemisphere our materials from the continents are less abundant than in the northern hemisphere, but observations from the ocean are very numerous.

128. *Three Systems of Winds.*—When we project all these observations upon a map of the earth, we find that the winds are naturally divided into three grand systems.
1. The equatorial system.
2. The winds of the middle latitudes; and,
3. The polar winds.

129. *The Trade Winds.*—Throughout nearly the entire equatorial region of the globe, whether over the land or on the ocean, the winds preserve a remarkable uniformity; on the northern side of the equator blowing almost invariably from some northeast quarter, and on the southern side of the equator from a southeast quarter. This system of currents is called the *trade winds.*

In the Atlantic Ocean, the N. E. trades extend on an average from about latitude 7° to latitude 29° N., while the S. E. trades extend to latitude 20° S. Between the N. E. and S. E. trades is a belt of calms or variable winds, extending at different seasons from 150 to 500 miles in breadth, and the centre of this belt is about five degrees north of the equator.

Throughout the northern half of the belt of the N. E. trades, the average direction of the winds is from N. 60° E.; but near latitude 10° they veer more to the east, and near their southern limit their direction is almost exactly east. The average direction of the S. E. trades is from S. 54° E.

The boundaries of the trade winds vary somewhat with the season of the year. During the summer they advance a few degrees farther toward the north, while in winter they recede somewhat toward the south. In spring, the centre of the belt of calms is only 1° or 2° north of the equator, while in summer it rises to latitude 9° or 10°.

130. *Winds in the Middle Latitudes.*—Beyond the borders of the trade winds in either hemisphere we find the prevalent winds at the earth's surface are from the west. In the northern hemisphere they blow from a point a little south of west, and in the southern hemisphere from a point a little north of west. This zone of westerly winds is from 25° to 30° in breadth; the westerly motion being most decided in the middle of the belt, but gradually diminishing as we approach the limit on either side. Throughout the middle latitudes of the United States, the average direction of the wind is from S. 80° W.; and the easterly winds are to the westerly in about the ratio of 2 to 5.

So, also, between the parallels of 40° and 60°, in the southern hemisphere, the prevalent direction of the surface-winds is from about N. 73° W.; and the easterly winds are to the westerly as 1 to 5.

131. *Direction of the Polar Winds.*—Beyond the parallel of 60° the general tendency of the winds is almost, without exception, toward the equator; but in some places the inclination is toward

Fig. 37.

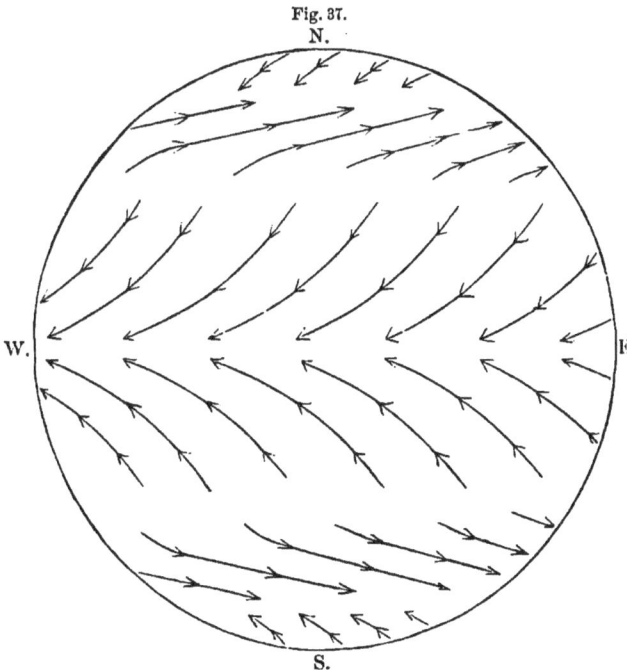

the west, and in others toward the east. In the northern hemisphere, beyond the parallel of 60°, northeast winds generally prevail, but in many districts the prevalent winds are from the northwest. Fig. 37 represents for every latitude the prevalent direction of the winds at the earth's surface.

132. *The Surface Winds.*—The winds here described are the winds which prevail at the earth's surface. They also extend to a considerable height, as is shown by observations on the summits of mountains, and by the observed direction of the clouds. It is believed that the directions here given are the average directions of the wind, as high as two miles from the earth's surface, and perhaps somewhat higher, including nearly (and perhaps fully) one half the weight of the entire atmosphere. Above this height we find an entirely different system of winds to prevail.

133. *Motion of the Upper half of the Atmosphere.*—It is evident that over any parallel of latitude, the northerly motion of the entire mass of the atmosphere must be exactly equal to its southerly motion, otherwise the atmosphere would be gradually withdrawn from certain portions of the earth, and would accumulate over other portions. If, then, in the equatorial regions, we find the average motion of the lower half of the atmosphere is *toward* the equator, the average motion of the upper half must be *from* the equator; and we actually find that in the northern hemisphere, within the region of the trade winds, the upper half of the atmosphere moves from the southwest. This is proved by the eruptions of volcanoes, and by observations on the summits of mountains.

134. *Evidence derived from Volcanoes.*—Within the limits of the trade winds are several volcanoes, which sometimes eject ashes to a great height, and these ashes indicate the direction of the stratum of air into which they rise. In the West Indies, in latitude 15°, on the island of St. Vincent, is a volcano which in 1812 emitted a vast quantity of ashes. A large mass of ashes fell upon the island of Barbadoes, which is ninety miles east of St. Vincent, although between the two islands the trade winds continually blow with such force that it is only by making a very long circuit that a ship can sail from the latter to the former. The ashes were doubtless transported by an upper current blowing in a direction contrary to that which prevailed at the surface of the sea.

A similar phenomenon was observed in January, 1835, on the great eruption of the volcano of Coseguina, in latitude 13° north, on the shores of the Pacific. Some of the ashes fell upon the island of Jamaica, at the distance of 700 miles in a direct line northeast from the volcano. At the same time, ashes were carried in the contrary direction westward, and fell upon a ship in the Pacific more than 1200 miles distant.

135. *Fine Dust transported by Winds.*—At several places in Southern Europe, Lyons, Genoa, etc., there has repeatedly fallen a fine dust, which was once supposed to come from the sandy plains of Africa; but Ehrenberg, by examination with the microscope, has shown that this dust contains microscopic organisms and dried infusoria. Among them he has found several South American species belonging to the valleys of the Oronoco and the Amazon, and which have not been found in any other part of the world.

We must then conclude either that this dust came in part from South America through the upper regions of the atmosphere, or these species exist in some other part of the world hitherto undiscovered. There is little doubt that the former is the true explanation, and we conclude that this dust from South America was elevated into the upper regions of the atmosphere, where it met a current from the southwest, in which it was transported a distance of over five thousand miles before it fell again to the earth.

136. *Winds on the Summits of Mountains.* — Observations on the summits of mountains indicate the same westerly current in the upper regions of the atmosphere. Upon the summit of Mauna Kea, on one of the Sandwich Islands, at the height of 13,951 feet, there is uniformly found a blustering wind from the southwest, while the regular trade wind from the northeast is blowing at its base.

The Peak of Teneriffe (12,205 feet in elevation) does not reach the limit of the lower half of the atmosphere, yet the wind here often blows from the southwest, and the clouds over the peak constantly move from the southwest in a direction opposite to the trade winds below. The traveler Bruce noticed a similar fact on the mountains of Abyssinia.

137. *Upper Current in the Middle Latitudes.*—Over the middle latitudes, at an elevation of about 10,000 feet above the earth's surface, we find a stratum of air uniformly moving from the northward. This is indicated by the following considerations:

1. In May, 1783, the famous volcano Hecla, in Iceland, commenced vomiting out smoke and ashes, which continued for a period of more than two months. This smoke rose to a great height in the atmosphere, and spread over nearly the whole of Europe, forming what was called a dry fog. It appeared first in the northwest part of Europe, gradually extending southward and eastward into Italy and even into Syria, which seems to indicate that during these two months there was an upper current of atmosphere moving from the northwest, all the way from Iceland to Syria.

During the same period a similar dry fog extended over a great part of North America, which seems to indicate the existence of another and probably higher current blowing steadily from the northeast.

During another eruption of this volcano in 1845, great quantities of the ashes fell on the Orkney Isles, and the ships navigating the neighboring seas.   •

2. Aeronauts who have ascended to the height of 10,000 feet in the middle latitudes, usually find the wind blowing from the west; and if they rise still higher, generally find the wind blowing from a point somewhat to the north of west.

3. Clouds chiefly prevail in the lower half of the atmosphere, and their average direction is about the same as that of the air at the earth's surface; but if, during a specially dry time, clouds are observed at a great elevation, they are generally found to move from a point north of west. According to six years' observations at Philadelphia, when the dew-point was 25° below the temperature of the air, the mean direction of the clouds was from N.55° W.

138. *Upper Current in the Polar Regions.*—It is evident that if in the polar regions the general progress of the surface current is toward the equator, there must be an upper current directed from the equator.

139. *Entire System of Atmospheric Circulation.*—We hence conclude that a section of the atmosphere made by a meridian

Fig. 38.

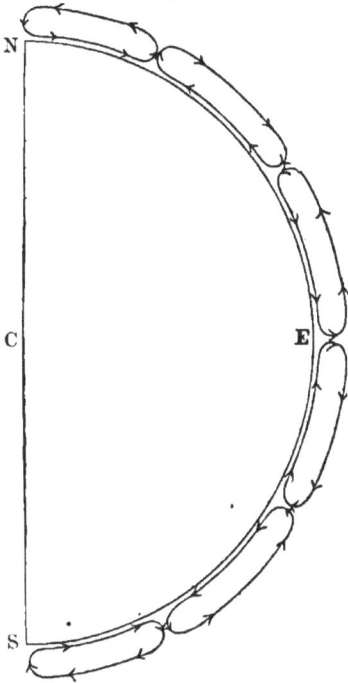

would exhibit the system of currents represented in Fig. 38, where N denotes the north pole, S the south pole, and E the equator. Within the tropics we find the surface current moving toward the equator, and the upper current from the equator. In the middle latitudes the surface current is moving from the equator, and the upper current toward the equator. In the polar regions the surface current is from the poles, and the upper current must therefore be toward the poles.

This diagram merely indicates whether the wind is moving to or from the equator. Its easterly or westerly motion could not be exhibited without a modification of the diagram. Throughout the equatorial belt of winds in the northern hemisphere the surface current is from the northeast and the upper current from the southwest; between the parallels of 30° and 60° the surface current is from the southwest and the upper current from the northward, while beyond the parallel of 60° the surface current is *toward* the equator, and the upper current is *from* the equator.

It is required to explain this system of atmospheric circulation.

140. *Causes of the Winds.*—There are three important causes which contribute to the production of wind.

1. Unequal atmospheric pressure.
2. Unequal specific gravity of the air; and,
3. The rotation of the earth.

Unequal pressure tends to produce motion in the atmosphere For conceive of two vertical columns of air extending to the top of the atmosphere, and imagine them to be connected near the earth by a horizontal tube. If the weight of one column exceeds, that of the other, the air must flow from the heavier to the lighter column, in the same manner as when water stands at unequal heights in the two arms of a recurved tube. The wind must therefore blow *from* places where the barometer is highest *toward* places where it is most depressed.

141. *Unequal Specific Gravity of the Air.* — Unequal specific gravity of the air may result from unequal *temperature* or from unequal *humidity.*

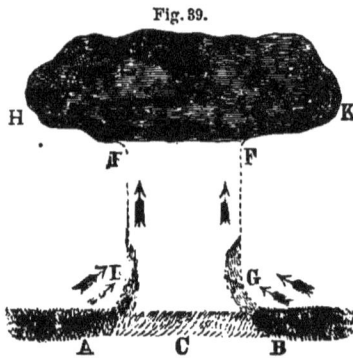

Fig. 39.

Let ACB, Fig. 39, represent an extended region of country, a portion of which, near C, is covered with sand, and becomes intensely heated by the rays of the sun, while at A and B the earth is covered with vegetation. The air which rests upon C, being more expanded than the surrounding air, rises, and its place is supplied by air flowing horizontally from A and B in the direction of the arrows. At the same time, the column of air, DEFG, being expanded, and rising above the surrounding atmosphere, overflows on each side in the direction of the arrows HK, producing upper currents moving in a direction contrary to the winds at A and B, and at a certain distance give rise to descending currents to supply the place of the air which near the earth's surface flows toward the heated region.

The motion here described may be illustrated by the following experiment: If in winter we partially open a door communicating between a hot and a cold room, and hold a lighted candle near the top of the crevice, the flame will be bent *outward* from the warm room, indicating a current of air from the hot to the cold room; but if we hold the candle near the **bottom** of the crevice, the flame will be bent *inward*, indicating a current from the cold to the hot room. We thus discover that the air flows out at the top of the heated room, while the cold air enters near the floor. In a similar manner, the unequal warmth of the earth's surface gives rise to currents of air of immense extent, the denser air flowing under and displacing the lighter.

The specific gravity of the vapor of water is only about two thirds that of dry air at the same temperature and pressure; and since it requires time for vapor to diffuse itself through the atmosphere, an excess of aqueous vapor must give rise to currents in the atmosphere in the same manner as inequality of temperature.

Even then, though the barometer may every where indicate the same pressure, the wind at the surface of the earth will tend from the colder to the warmer region, *from* the place where the atmosphere contains the least vapor *to* that where there is the most vapor.

142. *Mode of Propagation of Winds.*—The wind is first noticed near the heated column of air, and gradually extends to a greater distance from it. As the air moves from A and B toward the ascending column DEFG, the air at A and B is rarefied, and this rarefaction is communicated to the more distant air, and so on; that is, the wind is propagated in a direction contrary to that in which it blows. Winds thus propagated are called winds of *aspiration.* Winds which are propagated in the same direction as that in which they blow are called winds of *impulsion.* Examples of both of these classes of winds are found in all great storms, as will be shown in Chapter VI.

143. *Rotation of the Earth.*—The rotation of the earth would alone produce no permanent wind, because, if there were no other disturbing causes, the atmosphere would, by friction upon the earth's surface, soon acquire the same velocity of rotation as that of the portion of the earth upon which it rested; but the earth's rotation materially modifies the operation of other disturbing causes.

Since the earth is nearly a sphere, rotating upon its axis once in twenty-four hours, the velocity of rotation of different parallels of latitude is very different.

In latitude   0° the velocity eastward is 1036 miles per hour.
"   "   15° "   "   "   " 1000 "   "   "
"   "   30° "   "   "   "   897 "   "   "
"   "   45° "   "   "   "   732 "   "   "
"   "   60° "   "   "   "   518 "   "   "
"   "   75° "   "   "   "   268 "   "   "

F

144. *Relative Motion resulting from this Rotation.*—If a mass of quiescent air from the parallel of 30° could be suddenly transported to the parallel of 15°, it would have an easterly motion 103 miles per hour *less* than that of the parallel arrived at; that is, it would have a relative motion westward of 103 miles per hour. So also, if a mass of air from the parallel of 15° could be suddenly transported to the parallel of 30°, it would have an easterly motion 103 miles per hour *greater* than that of the parallel arrived at. That is, in general, if air is transported from the equator toward the poles, it will have a relative motion eastward; and if air is transferred from a higher latitude toward the equator, it will have a relative motion westward.

145. *Surface Winds in the Equatorial Regions.*—We have seen, Art. 20, that near the parallel of 32° the mean height of the barometer is greater than in any other part of the earth, and is .283 inch greater than it is near the equator. Also, the mean temperature of the surface air at the equator is about 12° higher than it is over the parallel of 32°. For both of these reasons, the air must tend from the parallel of 32° toward the equator; and if no other force acted upon it, the motion of the air in either hemisphere would be along a meridian toward the equator. But while the air from the parallel of 32° in the northern hemisphere flows toward the equator, it retains the easterly motion of the place from which it started, and in its progress southward reaches in succession parallels moving eastward more rapidly than itself. It therefore drags continually behind; that is, its motion with reference to the earth's surface is toward the west. Under the action of these two forces the progress of the air is toward the southwest, and the exact path described will depend upon the relative magnitude of the southerly and westerly motions.

A similar result must be produced on the south side of the equator, and thus originates a system of currents flowing from the northeast in the northern hemisphere, and from the southeast in the southern hemisphere.

146. *Upper Current in the Equatorial Regions.*—The mean temperature of the surface air at the equator is considerably higher than it is over the parallel of 32°, while near the upper limit of the atmosphere the temperature must be nearly the same in all latitudes. Now air is expanded by heat to the amount of $\frac{1}{491}$th part of its bulk for each degree of the thermometer. The atmosphere over the equator must therefore rise somewhat higher than it does over the parallel of 32°, notwithstanding the difference in the height of the barometer. If the earth were at rest, the air thus expanded at the equator would flow over at the top, and descend as along an inclined plane toward the middle latitudes. But while in the northern hemisphere an upper current flows toward the poles, it crosses in succession parallels of latitude whose easterly motion is less than its own; and since it retains the easterly motion which it had at the equator, it has a relative motion from the west, which, combined with the first northerly motion, carries it toward the northeast. Thus above the northeast trade winds we find an upper current moving from the southwest.

For a similar reason, in the southern hemisphere, above the southeast trades, the upper current moves from the northwest.

147. *The Surface Wind in the Middle Latitudes.*—Over the parallel of 32° the mean pressure of the air is .558 inch greater than over the parallel of 64°, and therefore at the earth's surface the air tends from the parallel of 32° toward the pole. The air in latitude 32° is indeed warmer, and therefore lighter than it is near the poles, and this creates a tendency of the surface current from the poles toward the equator; but the effect of the increased pressure of the air near the parallel of 32° is greater than that of its diminished density, and the air actually moves toward the poles.

But, while in the northern hemisphere the air from the parallel of 32° moves northward, it crosses successively parallels of latitude whose easterly motion is less than its own; and since it retains the easterly motion which it had at starting, it has a relative motion from the west, which, combined with the first northerly motion, carries it toward the northeast. Thus throughout the middle latitudes of the northern hemisphere the prevalent motion of the lower portion of the atmosphere is from the southwest, and, for like reasons, in the southern hemisphere the lower portion of the atmosphere moves from the northwest.

148. *The Surface Wind in the Polar Regions.*—It is believed that in the neighborhood of either pole the mean pressure of the atmosphere is somewhat greater than it is near the parallel of 64°, and, since the air is colder, it has a greater density. Both causes, therefore, conspire to impel the air toward the lower latitudes; and this force, combined with the effect of the earth's rotation, produces a northeast wind within the Arctic circle, and a southeast wind within the Antarctic circle.

149. *Ascending Current near the Parallel of* 64°.—We thus find that in the northern hemisphere the surface wind from each side of the parallel of 64° blows toward that parallel. This wind here rises from the earth's surface as it does near the equator, and becomes an upper current receding on either side from the parallel of 64°; but this upper current will not continue its course exactly in the direction of a meridian. As the air advances southward, it crosses successively parallels whose easterly motion is more and more rapid, so that, after some time, the direction of the upper current should be from the northeast. The northwest current, which seems generally to prevail at the height of two or three miles, may result from the partial mingling of this northeast current with the westerly wind which prevails near the earth's surface.

150. *Cause of the High Barometer near the Parallel of* 32°.—If the pressure of the barometer were the same at all points of the earth's surface, in consequence of the greater heat of the equatorial regions, there would be a general tendency of the surface air *toward* the equator, and of the upper air *from* the equator. This upper current could not, however, proceed on uninterruptedly to the pole, because the meridians converge, and their distance from each other continually diminishes, until they all meet at the poles. As the upper current of air recedes from the equator, it crosses successively parallels of less and less circumference, by which means the atmosphere is forced up to a corresponding height, and its pressure upon the earth's surface thereby increased. In latitude 32°, the distance between the meridians is nearly one sixth less than it is at the equator. This increased pressure of the air in the middle latitudes arrests the farther progress of the polar current, and a calm ensues. The upper air descends to the earth's surface, and joins the surface current toward the equator, where it again ascends, and thus maintains a perpetual circulation.

The high barometer near the parallel of 32° forces a surface current northward in opposition to the increased density of the air arising from a diminished temperature. Beyond the parallel of 64° the latter tendency is stronger than the former, and the surface current is from the poles.

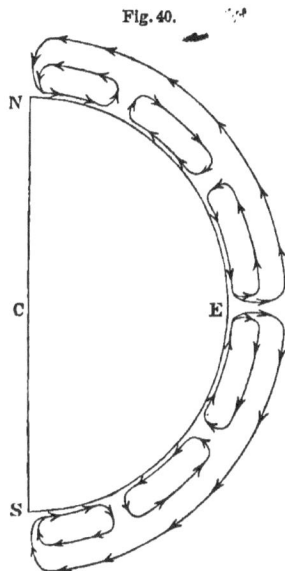

Fig. 40.

The cause of the low barometer near the parallel of 64° will be considered in Chapter VI., page 149.

There is some reason for supposing that in the most elevated regions of the atmosphere, where the atmosphere has nearly reached its limit of tenuity, the current over the equatorial regions, instead of descending to the earth's surface near latitude 32°, may continue on the same course over the middle latitudes to the polar regions, as represented in Figure 40; but the principal mass of the atmosphere is believed to circulate as represented in Fig. 38.

151. *The Monsoons.*—In mid-ocean the direction of the trade winds is quite uniform, but in the neighborhood of the continents great irregularities are observed. The most remarkable of these occur in the Indian Ocean, and are known by the name of monsoons. During the cooler half of the year, from October to March, the regular trade winds prevail here as in other parts of the northern hemisphere; but during the warmer half of the year, from April to September, the prevalent wind blows in the contrary direction, viz., from the southwest.

152. *Cause of the Monsoons.*—This change of wind results from the influence of the sun's heat upon the continent of Asia. In summer, the southern part of Asia becomes warmer than the Indian Ocean near the equator, and the cooler air from the ocean rushes northward toward the land to displace the heated air. This air, coming from a lower latitude, has an excess of motion toward the east, which, combined with the motion from the south due to the influence of heat, produces a wind from the southwest. This southwest wind sweeps over the high range of mountains north of Hindostan, by which means its vapor is condensed, forming excessive rains, by which means a vast amount of latent heat is liberated, and the surrounding air is still more expanded, thus adding to the force of the previous southwest current in a manner which is explained in Chapter VI., page 147.

During winter the Indian Ocean is warmer than the southern part of Asia, and the air from the land flows toward the equator, producing the usual northeast trade wind.

153. *Influence of the Seasons.*—Similar phenomena are noticed in every part of the world near the coasts of extensive continents. The continents being colder than the ocean during the winter and warmer during the summer, the winds tend from the land in winter and from the sea in summer. At most places this tendency serves simply to modify the direction of the prevalent winds. At Nantucket the winds blow from the northwest in winter and from the southwest in summer. Throughout the State of New York the average direction of the wind is 18° more southerly in summer than it is in winter. As we proceed southward this difference increases. At Washington the mean direction of the wind is northwest in winter and southwest in summer, while at many places on the coast of Florida the winds blow from the south in summer and from the north in winter, constituting well-marked monsoons. •

Similar phenomena are observed on the Pacific coast of the United States. At San Francisco, during winter, the winds blow most frequently from the northwest, while in summer they blow from the southwest with a steadiness equal to that of the trade winds within the tropics. At San Diego, lat. 32° 42′, throughout most of the year the wind blows steadily from the southwest, but during the winter months easterly winds are very prevalent, and sometimes for a month or two the average direction of the wind is from the northeast.

154. *Land and Sea Breezes.*—The diurnal change of temperature has a sensible effect upon the direction of the wind. This is seen in land and sea breezes which prevail on the coasts of continents and islands, particularly in tropical countries. During the day the land is heated more rapidly than the sea, and during the night it is more rapidly cooled. In the morning, the air in immediate contact with the land, being heated, is displaced by the cooler air in contact with the sea, and thus arises a breeze from the sea to the land. In summer this breeze usually springs up soon after 8 A.M., and attains its greatest intensity about the time of highest temperature. About sunset the breeze ceases entirely.

During the night the land becomes colder than the sea, and a breeze springs up from the land to the sea, which attains its greatest force about the time of lowest temperature. This breeze extends only to a short distance from the coast.

If no other cause operates to produce a wind, the direction of the land and sea breeze will be perpendicular to the coast; but if some other cause operates at the same time, the actual direction of the wind will be such as results from the composition of the two forces.

155. *Sea Breeze in the Temperate Zones.*—In the temperate zones the diurnal change of temperature produces a sensible effect in modifying the direction of the prevalent wind, and sometimes entirely reverses its direction. At New Haven the average direction of the wind throughout the year is 20° more southerly at noon than it is at sunrise, and from March to September the average change amounts to 35°. This effect is so uniform that sometimes every day, without exception for an entire month, the wind at noon is more southerly than it was at sunrise. Frequently the change amounts to 180°, the wind blowing from the north at sunrise and from the south at noon, and this phenomenon is rarely observed except during clear and pleasant weather, indicating that the change of wind is not the result of a great storm in progress.

156. *Temperature of the Wind.*—The temperature of the wind depends upon the quarter from which it blows and the countries which it has traversed. Generally in the northern hemisphere the winds from the south are warm, while those from the north are cold ; but the precise point of the horizon corresponding to the greatest heat and the greatest cold varies considerably. The following table shows how much the temperature of each wind is above or below the mean at New Haven, as deduced from a comparison of several years of observations.

Wind.	Temperature.	Wind.	Temperature.
N.	−2°.7	S.	+3°.2
N.E.	−0 .6	S.W.	+4 .0
E	+0 .5	W.	−1 .1
S.E.	+1 .2	N.W.	−4 .5

If we represent these differences of temperature by the ordinates of a curve, we shall obtain a
diagram like Fig. 41, which indicates
that at New Haven the highest temperature accompanies a wind from
south 33° west, and the lowest temperature corresponds to a wind from
north 40° west, the mean difference in the temperature of these
two winds being 8°.7.

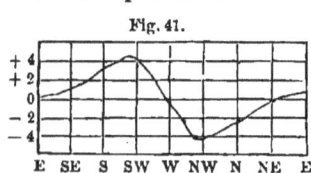

Fig. 41.

In many parts of Europe the coldest wind comes from a quarter somewhat east of north, and the hottest wind generally comes
from a point a little west of south.

157. Hot Winds of Deserts.—On the deserts of Africa and Arabia there sometimes prevails a wind extremely dry and intensely
hot, which raises clouds of sand, and transports it to a great distance. These winds are known by the name of *simoon, harmat-
tan*, etc., according to their locality. Plants are withered by this
wind; men and animals suffer intensely from the heat and dryness of the air; and entire caravans have been buried in the drifting sand. This dust is sometimes transported across the Medi-
terranean into Spain, Sicily, and Italy, where the wind which
brings it is known by the name of *sirocco*. In Sicily, during its
continuance, the thermometer sometimes rises to 110 degrees in
the shade.

158. Cold Winds from Mountains.—In mountainous countries
the winds from certain quarters are celebrated for their low temperature, and frequently for their dryness. The westerly winds
which cross the range of the Rocky Mountains, being cooled by
elevation, deposit most of their moisture upon the west side of
the mountain, and when they descend upon the eastern side they
are cold and dry winds. Hence along the eastern margin of the
Rocky Mountains rain seldom falls, and ordinary agricultural
products can not be raised without artificial irrigation.

The high mountains of South America produce effects still more remarkable. In Peru, between two great chains of the Andes, Fig. 42, in latitude 16° S., at the height of 13,000 feet, is a desolate table-land called the Punos, extending about 500 miles in length by 100 in breadth. The trade wind, by passing over the eastern chain of mountains, is reduced to a very low temperature, and nearly all its vapor is condensed in the form of rain or snow. When the air descends upon the western side of the mountain it is so cold and dry that the bodies of dead animals exposed to it are dried up like mummies, without any signs of putrefaction. Prescott states that the ancient Peruvians preserved the bodies of their dead for ages by simply exposing them to the cold dry air of the mountain.

Fig. 42.

PUNO.

<hr />

CHAPTER V.

PRECIPITATION OF THE VAPOR OF THE AIR.

SECTION I.

• DEW.

159. *Effect of Radiation of Heat.*—All bodies on the surface of the earth send out rays of heat toward the sky, and when they radiate more heat than they receive, their temperature falls below that of the surrounding air. In order to study these effects, we place a number of thermometers upon the ground on substances of different kinds, and suspend other thermometers in the air at various elevations, and compare the readings of these thermometers simultaneously at all hours of the day and night. Very careful observations of this kind were made at Greenwich, England, for two years, and it was found that a thermometer placed on grass fully exposed to the sky frequently sinks ten degrees below a thermometer suspended four feet from the ground. On nine nights the difference of temperature was more than 15°, and in one instance a thermometer placed on raw wool sunk 28°.5 below one suspended eight feet from the ground.

Radiation of heat from the earth to the sky takes place at all times, both day and night, and in all states of the sky. Generally, when the sun is above the horizon, the heat received from it by the earth exceeds that which is radiated from the earth. Sometimes, however, in places sheltered from the sun, but open to a considerable portion of the sky, the amount of heat radiated exceeds that received from the sun and all other sources, so that grass may continue colder than the air during the day as well as the night. This difference at midday has been known to amount to ten degrees.

160. *Effect of Partial Exposure to the Sky.*—Whatever impairs the free exposure of an object to the sky causes its temperature to decrease less than it would if the exposure was complete. This effect is produced by spreading a sheet of cloth over the ground, even though it be at a considerable elevation. The thinnest cambric handkerchief produces a decided effect. Trees and buildings, and whatever conceals a part of the sky, diminish the effect of the radiation of heat from the surface of the earth.

Clouds produce the same effect as an artificial covering. From an average of all the Greenwich observations, it was found that a thermometer placed on grass fully exposed to the sky sunk below a thermometer suspended four feet from the ground as follows:

On cloudless nights,   9.3 degrees.
" nights half cloudy,   7.3 "
" " principally cloudy, 6.8 "
" " entirely cloudy, 3.4 "

161. *Radiation from different Substances.*—Thermometers placed on different substances exhibit very unequal reduction of temperature on the same night. When a thermometer placed on grass sinks ten degrees below one suspended four feet from the ground, a thermometer placed on raw wool will sink 12° or 15°; a thermometer placed on copper will sink 8°; on paper, 6°; and on brick, only 3° or 4°. Tab. XXXII. shows the average results found for a great variety of substances. These numbers indicate the comparative radiating power of different substances for heat.

162. *Increase of Temperature with Elevation.*—By suspending thermometers at different elevations above the earth from one or two inches up to 200 feet, it is found that the loss of heat by nocturnal radiation is quite sensible at the elevation of 50 feet, and does not entirely disappear at the height of 150 feet. During the night, therefore, the temperature of the air *increases* as we rise above the earth's surface. In England, according to the average of observations continued throughout the year, if a thermometer placed on grass fully exposed to the sky be taken as the zero, a thermometer one inch above it would read 3° higher;

"	six inches	"	"	6°	"
"	one foot	"	"	7°	"
"	twelve feet	"	"	8°	"
"	fifty feet	"	"	10°	"
"	one hundred and fifty feet			12°	"

and the effect is appreciable at still greater elevations.

163. *Origin of Dew.*—When, in consequence of radiation, objects near the earth's surface, such as grass and leaves of vegetables, become cooled below the dew-point in the vicinity, they condense upon themselves a portion of the vapor which is present in the atmosphere, in the manner explained in Art. 99, and this moisture is called *dew.* The amount of dew thus deposited is greatest upon those substances whose temperature is the lowest, being proportional to the amount of depression of their temperature below that of the dew-point. Dew, therefore, does not *fall* from the sky like drops of rain, as was formerly supposed, but the vapor of the air is condensed by coming in contact with the cold surface of the object upon which the dew collects.

In some parts of the world, nearly all the moisture which the earth ever receives comes in the form of dew. This is particularly true of some parts of Egypt and Arabia.

164. *Circumstances favorable to Dew.*—The circumstances most favorable to the production of dew are mainly those which are most favorable to the loss of heat by radiation. These are,

1st. A cloudless night and unobstructed exposure to the sky. The deposition of dew is immediately checked by clouds which reflect back to the earth the heat radiated from it. The same effect is produced by any artificial covering, even though of the thinnest texture. Hence, also, plants placed beneath a tree or near a building collect much less dew than those which are freely exposed to the sky.

2d. A nearly tranquil atmosphere. A slight breeze, by renewing the air which has deposited its excess of vapor, renders the dew more abundant; but a fresh breeze, by agitation of the air, produces a mingling of the air at different elevations, equalizing the temperature throughout, so that the air at the earth's surface can not become much colder than the superincumbent atmosphere. Little dew is therefore deposited on windy nights.

3d. *A moist atmosphere.* When the atmosphere is most humid, a given reduction of temperature will sooner reach the dew-point, at which the deposition of moisture begins. An abundant dew is regarded as an indication of approaching rain, because it proves that the air contains a large quantity of vapor.

4th. Good radiators and bad conductors of heat are required for collecting the dew. Different substances, having the same exposure, do not collect the same amount of dew. Wool radiates heat freely, and, being a bad conductor, collects a large amount of dew; while but little dew is deposited on polished metals, since they are good conductors of heat, and must be reduced throughout to a low temperature before any dew can be deposited upon them. If similar plates of polished glass and steel are exposed alike upon the ground during a favorable night, in the morning the glass will be drenched with dew, while the brightness of the metal will be scarcely dimmed. The glass radiates heat more rapidly than the metal, and, being a bad conductor, draws but little warmth from the earth to supply its loss; while the metal, being a good conductor, readily derives heat from the warm soil below.

165. *Dew during the Day.*—The deposition of dew sometimes commences before sunset. It continues at all hours of the night, provided the weather remains favorable; but more dew is formed after midnight than before; and the deposition sometimes continues after sunrise.

In places sheltered from the sun, but open to a considerable portion of the sky, dew is sometimes deposited on grass even at midday.

166. *Where there is no Dew.*—Dew is not deposited on the surface of large bodies of water whose temperature is above 40°, for as soon as the particles at the surface are cooled they become heavier and sink, while warmer and lighter particles rise to the top, by which means the surface of the water is maintained at nearly the same temperature as the surrounding air.

In the midst of sandy deserts, on account of the dryness of the atmosphere, dew is almost entirely unknown. Travelers upon the deserts of Africa and Asia are notified of their proximity to lakes or rivers by the appearance of dew.

But little dew is deposited in cities, because most of the objects there found are poorer radiators than the leaves of vegetables, and because the heat of the city is always greater than that of the surrounding country.

167. *Amount of Dew determined.*—Attempts have been made to determine the total amount of dew annually deposited in different countries. This is sometimes done by exposing a plate of glass or some other substance to the sky, and carefully weighing the amount of moisture deposited upon it. In this way it has been concluded that in Italy and the south of France the annual deposit of dew amounts to a little more than a quarter of an inch. Such results, however, are not very reliable, since they are greatly influenced by the radiating power of the plate employed, and also by its position.

SECTION II.

HOAR-FROST.

168. *Formation of Hoar-frost.*—Hoar-frost is formed under the same circumstances as dew, with the exception of a lower temperature. When the temperature of plants falls below 32° the moisture of the air is condensed upon them in the solid state, and forms a layer of spongy ice. Hoar-frost, therefore, is not frozen dew, but the moisture of the air is deposited in the solid form, without having passed through the liquid condition. Hoar-frost, like dew, is deposited chiefly upon those bodies which radiate best, such as plants and the leaves of vegetables, and the deposit is made principally on those parts which are turned toward the sky.

Since plants sometimes become cooled by radiation from 12° to 15° below the temperature of the surrounding air, a frost may occur although a thermometer a few feet above the ground does not sink to 32°. During a clear and still night, when a thermometer six feet above the ground sinks to 36°, a very heavy frost may be expected, and a slight frost may occur when the same thermometer sinks only to 47°.

169. *How Plants are protected from Frost.*—Whatever prevents the radiation of heat serves also to check the formation of hoar-frost. During the cold nights of spring, plants which are sheltered by trees are less liable to be injured by frost than those which are fully exposed, and a thin covering of cloth or straw will generally afford entire protection. A garden may frequently be saved from injury by kindling a small fire, which shall envelop the plants in a cloud of smoke. Fogs and clouds also protect vegetation from the effects of frost.

170. *Frost in Valleys.*—Plants are often killed by frost in the valleys and up to a certain height upon the hills, while above this limit.they entirely escape injury. It has been found by observation that a thermometer attached to a high tower in a valley indicates at night the same average temperature as a thermometer on the side of a neighboring hill upon the same level. This indicates that during a tranquil night the cold air resulting from radiation at the surface of the earth settles in the valleys in consequence of its greater density, and the warm and cold air are arranged in nearly horizontal strata like liquids of different densities.

171. *Crystalline Structure of Hoar-frost.*—Hoar-frost generally exhibits a crystalline structure, and consists of long spiculæ, which are found to be hexagonal prisms with angles of 120°. These spiculæ are frequently seen in great perfection in the frost which forms on wooden fences, on the decayed branches of trees, etc.

Fig. 43.   Fig. 44.

When a thin film of water freezes upon a flat surface of glass or stone, it often forms a great variety of beautiful figures, sometimes resembling the leaves of certain plants, the leaves of the palm-tree, or the feathers of birds, Figs.

43 and 44. In cold weather, smooth and flat stones upon the side-walk are often covered with these figures, which, upon examination, are found to consist mainly of spiculæ more or less perfectly formed.

A species of hoar-frost occurs when a warm wind succeeds a period of severe cold weather. Stone buildings are then often covered with an incrustation of minute crystals caused by the low temperature of the stone, which condenses and congeals the moisture of the air.

SECTION III.

FOG.

172. *Condensation of the Vapor of the Atmosphere.*—The vapor in the atmosphere is nearly or quite transparent; but when, from any cause, the air becomes cooled below the dew-point, a portion of its vapor is precipitated in the form of drops of water extremely minute, which affect the transparency of the air, and form *fog* or *cloud* according as it occurs near the surface of the earth, or in the upper regions of the atmosphere. If we compress moist air in a close vessel and allow it suddenly to escape, the air, by its expansion, will be cooled, and a slight fog be produced; but the air soon regains its warmth, the drops of water return to the state of vapor, and the fog is dissipated.

When steam rises from a vessel of warm water and mingles with a cold atmosphere, a portion of the vapor is condensed and a mist is formed. This mist is sometimes, but improperly, called vapor. *Vapor of water* is a gaseous body, while *mist* is a liquid body. A similar condensation often takes place in nature upon a large scale, and the mist is then called a fog.

173. *Fogs over Rivers in Summer.*—At certain seasons of the year, especially during the latter part of summer, upon nearly every clear and still night, fogs form over rivers and lakes. At night the temperature of the air over the land becomes cooler than the water of lakes and rivers. The vapor which rises at such a time from the warm water is condensed by contact with the cooler air from the land, and a fog is formed, which seems to rest upon the water.

On a clear and quiet morning in the month of August, an observer on the summit of Mount Washington sees the bed of the

Connecticut River distinctly traced by a long line of fog, and the position of a multitude of surrounding lakes is indicated in the same manner, while other portions of the country are entirely free from fog.

That this fog is formed by the vapor of the warm water rising into an atmosphere which is cooler than the water, is proved by observations of the thermometer. On a morning in July, when the Connecticut River was covered with a thick fog, the temperature of the water was found to be 73°, while the temperature of the air over the neighboring land was only 68°.

Such fogs generally disappear soon after sunrise. Sometimes, from the effect of the sun's heat, they are seen to ascend and rise above the hills, forming clouds, which soon disappear with the increasing heat of the sun.

Fogs are often formed in a similar manner over harbors, bays, etc., and these fogs, by a gentle current, are often drifted over the land. In this manner a sea fog sometimes spreads over the city of New York, and extends several miles up the Hudson River.

174. *Fogs in Spring and Winter.* — During the spring of the year, fogs are sometimes formed over rivers, when the temperature of the water is colder than that of the surrounding air. In this case the warm and moist air of the neighboring land is chilled by coming in contact with the cold water, and a portion of its vapor is condensed.

In the same manner, after a warm rain in mid-winter, dense fogs are sometimes formed by a warm and moist air flowing over a country which is covered with snow; or the fog may result from the moist air becoming cooled by contact with a frozen soil. Indeed, a fog may be formed at any time at a distance from large bodies of water, when the vapor which rises from a very moist soil mixes with a cold atmosphere.

In the same manner, fogs are often formed on the sides of mountains. The warm air from the valleys being forced up the sides of the mountain, its vapor is condensed, partly by the cold of elevation, and partly by contact with the cold surface of the mountain.

175. *Where Fogs are most Prevalent.*—On the Atlantic Ocean, from 30° south to 35° north latitude, fogs are almost unknown.

On the northern side of the Gulf Stream they are of common occurrence, but they are most prevalent near the Banks of New-foundland. These fogs occur in every month of the year, but they are most prevalent in summer, when the Banks are enveloped in fog nearly half the time. The vapor which causes these fogs is furnished by the warm air of the Gulf Stream, and it is condensed by the cold air of the Banks, the contrast of temperature being here more sudden than is found in any other part of the Atlantic Ocean. During the month of July the water on the Banks frequently has a temperature of 45°, while within a distance of less than 300 miles the Gulf Stream has a temperature of 78°. The contrast of temperature is almost equally great in January, but fogs are less frequent in winter, because at that period the air is more agitated by storms, which tend to equalize the temperature over different parts of the ocean.

In the South Atlantic, beyond the parallel of 30°, fogs are of common occurrence, but they are nowhere so prevalent as on the Banks of Newfoundland.

176. *Fogs of Polar Regions, etc.*—Fogs are very prevalent in the Arctic regions, particularly in summer. During an Arctic summer the temperature of the earth rises much above that of the sea, portions of which are covered with immense fields of ice. The air resting upon the earth partakes of its temperature, and when this warmer air is brought in contact with the colder ocean, a portion of its vapor is condensed, and a heavy mist is formed.

During winter, England and the neighboring portions of the Continent are frequently enveloped in dense fogs, and in those towns where bituminous coal is used abundantly the sky is sometimes so darkened by the mixture of fog and smoke that locomotion even at midday becomes almost impossible. In London, during winter, the streets are sometimes lighted with gas all day, and travel through the city is attended with serious danger. This fog results from the warm air of the sea spreading over the cold land.

177. *Where Fogs do not Prevail.*—Fogs are never formed when the air is very dry, and therefore they are never known in deserts. Fogs are not common in tropical countries except in the neigh-

G

borhood of mountains; but the summits of mountains, even under the equator, are habitually shrouded in fog or cloud.

178. *The Vesicular Theory.*—Since fog consists of particles of a
liquid which is nearly eight hundred times denser than the air,
it has been thought difficult to explain how it can be sustained in
the atmosphere. Some have supposed that the particles of fog
are hollow, each consisting of a sphere of air surrounded by a
thin envelope of water like a soap-bubble. Such a hollow sphere
is called a *vesicle*, and this theory of the constitution of fog is called the *vesicular theory*.

179. *Argument from the appearance of Mist.*—Some observers
who have examined with a magnifying-glass the particles of mist
rising from hot water, have detected on their surface colored rings
like those seen on soap-bubbles, indicating that their structure
was analogous to that of soap-bubbles, and it has been inferred
that the particles of fog generally have a similar constitution.
Water ordinarily contains minute bubbles of air, and when the
water is warmed these bubbles expand and rise to the surface.
They often rest upon the surface of warm water, and, being surrounded by a thin film of water, they should exhibit colored rings
like soap-bubbles, but there is no evidence that the particles of
fog are generally so constituted. A fog is formed from the vapor
of water previously existing in the air in the gaseous state, and
when this vapor returns to the liquid condition there is no evidence that it assumes the form of hollow spheres.

180. *Argument from the absence of a Rainbow.*—It is contended
that the particles of fog can not be solid spheres of water, because
if they were, a rainbow should be seen whenever the spectator is
situated between the sun and a fog. A rainbow is formed by the
reflection of the sun's rays from falling drops of water. These
drops are spheres of water; and since a fog does not form a rainbow, it is contended that the particles of the fog can not be solid
drops. But it has been shown that when the spheres of water
are very small, as is the case with a fog, a bow should indeed be
formed, but the different colored bands are very broad, and their
light is proportionally feeble. Moreover, if the spheres are not
all sensibly of the same diameter, there will be formed simultane-

ously bands of different breadths, which will be superposed upon each other in such a manner that the different colors will be very much blended, and produce a light which is nearly white, forming thus a very faint and nearly white rainbow, or rather *fog-bow*, which corresponds exactly with the facts. When a spectator is situated between the sun and a bank of fog, a white bow is often seen with but little appearance of prismatic colors, and the breadth of the bow is about double that of an ordinary rainbow.

Thus the absence of the common rainbow in fogs not only does not establish the vesicular theory, but the existence of the white fog-bow positively refutes this theory.

181. *Argument from the Constitution of Clouds.*—Fogs evidently have the same constitution as clouds. Now, when clouds are formed at a low temperature, their particles are solid, consisting of spiculæ of ice, which, united, form flakes of snow. But we find nothing of the vesicular constitution in snow-flakes; nevertheless, clouds composed of spiculæ of ice remain suspended in the air for hours, and sometimes days in succession. The vesicular hypothesis, therefore, is not necessary to account for the permanence of clouds and fogs, and there is no evidence that they are ever thus constituted.

182. *How Fog is sustained in the Air.*—The particles of fog are sustained in the air in the same manner as a cloud of dust is sustained. A cloud of dust remains for a long time suspended in the air, although each particle of dust may consist of matter two thousand times as dense as the air in which it floats. When the air is perfectly tranquil these particles do indeed fall, but they descend so slowly that their motion is only perceptible after the lapse of a considerable interval of time.

183. *Diameter of Particles of Fog.*—The diameter of particles of fog is very variable, being sometimes so small that the individual particles can not be separately seen, and it is only in *mass* that they make any impression upon the eye, and they are found increasing in size until they fall with considerable velocity, when they are called rain-drops.

The diameter of the smallest visible particles of fog has been estimated at $\frac{1}{1800}$ inch; and when the diameter of the particles becomes equal to $\frac{1}{80}$ inch, they fall with an appreciable velocity, and are called rain-drops.

184. *Indian Summer.*—At certain seasons of the year there occurs a peculiar phenomenon called a *dry fog.* In the United States this frequently occurs in November, or the latter part of October, and this period is commonly known by the name of *Indian Summer.* This period is characterized by a hazy condition of the atmosphere, a redness of the sky, absence of rain, and a mild temperature. This appears to result from a dry and stagnant state of the atmosphere, during which the air becomes filled with dust and smoke arising from numerous fires, by which its transparency is greatly impaired. A heavy rain washes out these impurities and effectually clears the sky.

This phenomenon is not peculiar to the United States, a similar condition of the atmosphere being frequently observed in Central Europe. Moreover, this dry and stagnant state of the atmosphere is not limited to a single season of the year. The long periods of drought which frequently prevail in summer are characterized by a like condition of the atmosphere.

185. *Prevalence of Volcanic Ashes, etc.*—Sometimes a dry fog continues for several weeks, and prevails over a vast area, exhibiting very peculiar characteristics. These fogs have been ascribed to the presence of fine volcanic ashes in the atmosphere, and perhaps also of substances foreign to the earth.

In 1783 such a fog prevailed over all Europe, and continued for more than a month. It was preceded by a remarkable eruption of the volcano Hecla, in Iceland, which for a long time emitted smoke of unusual density.

In 1831 a similar fog prevailed in the United States, in Europe, and on the coast of Africa. It obscured the air to such an extent that the sun could be observed all day with the naked eye, without the interposition of any colored glass. At night the fog seemed decidedly phosphorescent, and emitted an appreciable amount of light, which could not be ascribed to the reflected light of the stars.

SECTION IV.
CLOUDS.

186. Clouds differ from fogs only in their elevation above the earth. A fog resting on the top of a mountain is called a cloud. A cloud resting on the surface of the earth is called a fog.

187. *Classification of Clouds — Cirrus.* — Clouds present an infinite variety of forms, yet they may be divided into six classes, each presenting characteristics tolerably distinct. Three of these modifications are primary, and three are compound.

The *cirrus* cloud consists of long, slender filaments, either parallel or diverging from each other, and often presents the appearance of a lock of cotton whose fibres are electrified so as powerfully to repel each other. These clouds appear to have the least density, the greatest elevation, and the greatest variety of form. They are generally the first to make their appearance after a period of perfectly clear weather. Indeed, in fair weather, the sky is seldom entirely free from small groups of cirrus clouds. They are believed to be composed of spiculæ of ice or flakes of snow, floating at a great height in the air. At the height at which they prevail, the temperature of the air is below 32° even in midsummer. It is among clouds of this variety that halos and parhelia are formed, phenomena which are ascribed to the refraction of light by minute prisms of ice.

188. *Cumulus.* — The cumulus cloud usually consists of a hemispherical or convex mass, rising from a horizontal base. It is much denser than the cirrus, and is formed in the lower regions of the atmosphere. In fair weather the cumulus often forms a few hours after sunrise, goes on increasing until the hottest part of the day, and disappears about sunset. We often see near the horizon large masses of cumulus clouds, which resemble lofty mountains covered with snow.

The rounded top of the cumulus results from the mode of its formation. When the surface of the earth is heated by the rays of the sun, currents of warm air ascend, and as soon as they reach a certain height a portion of their vapor is condensed, and forms cloud; and since the upward motion is greatest under the centre of the cloud, the vapor is there carried up to the greatest height.

In like manner, when steam escapes in large quantities from the boiler of a steam-engine, especially in a damp atmosphere, it forms a rounded mass of mist.

189. *Stratus.*—The stratus cloud is a widely-extended, continuous, horizontal sheet, often covering the entire sky with a nearly uniform veil. This is the lowest of the clouds, and sometimes descends to the earth's surface.

190. *Compound Modifications.*—The *cirro-cumulus* consists of small, well-defined, rounded masses, in close proximity. These little rounded clouds, on account of their fleecy appearance, are sometimes called woolly clouds. The cirro-cumulus is frequent in summer, and is attendant on warm and dry weather.

The *cirro-stratus* consists of delicate fibrous clouds spread out in strata, which are either horizontal, or but slightly inclined to the horizon. This cloud appears to result from the subsidence of the fibres of the cirrus to a horizontal position. Sometimes the whole sky is so mottled with this kind of cloud as to resemble the back of a mackerel, and is hence called the *mackerel-sky*. The cirro-stratus precedes wind and rain, and is almost always to be seen in the intervals of storms.

The *cumulo-stratus* consists of the cumulus blended with the stratus, and is formed in the interval between the first appearance of the fleecy cumulus and the commencement of rain. On the approach of a thunder-storm the cumulo-stratus clouds are often seen in great magnificence, and present those peculiar forms known in some places by the name of *thunder-heads*. All these varieties of cloud are represented in Plate II.

191. *Best Mode of observing the Clouds.*—In order to be able to distinguish well the form of clouds, it is often necessary to diminish their brilliancy by viewing them through a glass of a deep blue color, or by reflection from a mirror of black glass. We are thus able to detect peculiarities which entirely escape observations with the unassisted eye.

The appearance of a cloud often changes greatly with its change of position in the heavens. The peculiarities of clouds are generally more noticeable when they are near the zenith than when they are near the horizon.

Besides the six modifications of clouds above enumerated, Howard admitted a seventh, which he called the cumulo-cirro-stratus, or *nimbus*, to denote a cloud or system of clouds from which rain is falling; but it is often so difficult to distinguish between the stratus and nimbus that it seems inexpedient to retain the last division.

192. *Average degree of Cloudiness.*—Clouds are more prevalent in some parts of the world than in others. Throughout New England, on an average for the whole year, $\frac{53}{100}$ths of the sky are covered with clouds, while in the Southern States the average is $\frac{47}{100}$ths. Near the equator, between the N. E. and S. E. trade winds, there are places where the sky is almost constantly covered with clouds. At St. Helena, at an elevation of 1764 feet, on an average for the whole year, $\frac{89}{100}$ths of the sky are covered with clouds, and on the tops of high mountains the sky is seldom free from clouds.

Throughout most of Great Britain the average cloudiness is $\frac{70}{100}$ths, while at Bombay it is only $\frac{36}{100}$ths, and at Sacramento, California, it is only $\frac{31}{100}$ths.

193. *Height of Clouds.*—The height of a cloud may sometimes be measured in the same manner as the height of any other inaccessible object, by simultaneous observations of its direction at two stations. More satisfactory results may, however, be obtained by ascending in a balloon, and noting the height of the barometer at the instant of entering a cloud, and again when emerging from it; the barometer affording the means of computing the corresponding altitudes. In mountainous countries we may sometimes determine the height of a cloud by comparing it with some peak of known elevation near which the cloud is carried by the wind.

The height of clouds is very variable, and their mean elevation is not the same in different countries. The stratus cloud often descends to the earth's surface. In pleasant weather, the lower limit of cumulus clouds varies from 3000 to 5000 feet elevation, and their upper limit from 5000 to 12,000 feet. Cirrus clouds are never seen below the summit of Mount Blanc, which has an elevation of 15,744 feet.

Clouds are sometimes seen above the summit of Chimborazo,

which has an elevation of 21,424 feet. Gay-Lussac and Glaisher, in their different balloon ascents to the height of 23,000 feet, saw cirrus clouds which appeared considerably above them. It is estimated that the greatest height at which visible clouds ever exist does not exceed ten miles.

194. *Vertical Thickness of Clouds.*—The vertical thickness of clouds does not generally exceed half a mile, but cumulus clouds are sometimes formed of enormous magnitude and height. It has been computed that the tops of cumulus clouds sometimes attain the height of four miles, while their bases are not more than half a mile above the earth's surface.

195. *Formation of Clouds.*—The vapor generated at the surface of the earth by the heat of the sun tends, by its expansive force to spread in all directions, but especially upward, and forms an atmosphere of vapor whose density decreases with the elevation. Since the temperature of the air sinks as we rise above the earth, it may happen that at a certain height the tension of the vapor is greater than corresponds to the temperature which prevails at that elevation, in which case a portion of the vapor will be precipitated and form a cloud.

Vapor may also be transported to a great height by the ascending currents of air caused by the heat of the sun. These currents ordinarily give rise to *cumulus* clouds, and it frequently happens that the sky, though clear in the morning, is filled with those clouds at noon.

Any cause which chills a humid air determines the formation of cloud. A cold wind penetrating a humid air, or a warm and humid wind penetrating a cold air, causes the precipitation of vapor and the formation of cloud. At the close of a warm day, especially after a rain, clouds are frequently formed, which increase during the night, and are dissipated the next day by the effect of the sun's heat.

196. *Summits of Mountains enveloped in Cloud.*—The summits of high mountains are almost always enveloped in clouds, even though every other portion of the sky is perfectly clear. This is not due to any attraction between the mountain and the cloud, but rather the mountain *causes* the cloud. The effect of an interposed mountain is to force a horizontal wind up to an unusual height where the temperature is low, and when the temperature of the air is reduced below its dew-point, a portion of its vapor must be condensed and form cloud. Thus, let ABC be a mountain interposed in the path of a horizontal current of air. The air is by this means forced upward, and made to glide along the side of the mountain. If DE represents the height at which the temperature of the air is just equal to the dew-point of this current, then, as soon as the wind passes above the line DE, a portion of its vapor will be condensed, and a cloud will be formed which will envelop the summit of the mountain. But when the air, descending from the mountain on the opposite side, passes below the line DE, it attains a temperature which is above the dew-point, and the cloud is redissolved.

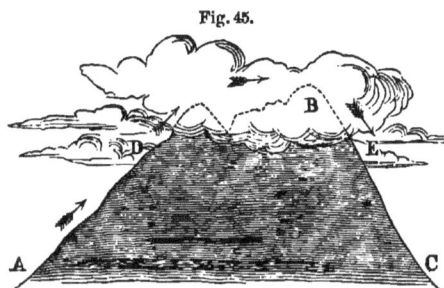

Fig. 45.

It is sometimes thought strange that the strong wind which usually prevails on the summits of mountains does not *blow away* the cloud. Undoubtedly the cloud is drifted off with the wind, but its place is instantly supplied with new cloud. Thus, although the cloud on the summit of the mountain appears perfectly stationary, the particles which compose the cloud are continually changing. A somewhat similar effect often takes place over countries which are tolerably level. The sky does not become overcast solely from clouds which are drifted by the wind from places beyond the horizon; but new clouds frequently form directly in sight of an observer. On the contrary, a cloudy sky sometimes clears up, not because the clouds are drifted off by the wind, but because they are converted into vapor by the increasing heat of the air.

197. *How Clouds are Sustained.*—Since clouds consist of particles which are heavier than the surrounding air, they must sink, even though it be slowly, and we might conclude that in calm weather they must at length fall to the ground. The particles of a cloud, however, in pleasant weather can not reach the ground, because in descending they meet a warmer stratum of air which is not saturated with vapor, when the lower part is again converted into vapor and disappears. This explains why the base of the cumulus cloud is uniformly horizontal. While, however, the particles on the lower side of the cloud are dissolved, the upper part of the cloud is continually increasing by the condensation of new vapor, which is carried upward by ascending currents of air, by which means the cloud appears to maintain a constant elevation above the earth.

198. *Currents in the Air.*—Observations of the clouds often disclose the existence of currents in the atmosphere flowing in various and perhaps opposite directions. We sometimes notice a stratum of clouds moving nearly in the direction of the air at the earth's surface, while at a greater elevation we observe a stratum moving in a different direction, and sometimes a third and perhaps a fourth moving in still other directions. Such cases are of frequent occurrence near the commencement or during the progress of a great storm.

199. *Peculiar Arrangement of Clouds.*—Clouds sometimes assume remarkable forms, which we can not ascribe to chance. Sometimes cirro-cumulus clouds arrange themselves in long lines,

Fig. 46.

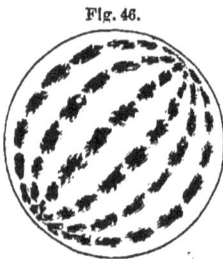

stretching quite across the horizon. Sometimes several such lines stretch across the sky in nearly parallel directions, while occasionally the whole heavens are covered with such bands, which seem to diverge from one point of the horizon, and converge to the opposite point. Such bands generally point from southwest to northeast, as shown in Fig. 46.

The apparent curvature of the lines is the effect of perspective, the bands being in fact parallel to each other. The direction of these lines generally coincides with that of the wind, and it has been suspected that these lines of cloud serve as conductors of currents of electricity, and this may be the agent which causes the clouds to assume such artificial forms.

200. *Shadows of Clouds.*—When the atmosphere is filled with a

dense haze, the shadows of houses and trees are often distinctly depicted upon the haze. So also when the sky is somewhat hazy, the shadows of clouds can be distinctly traced in the sky by dark lines proceeding from the sun. Such a haze most frequently prevails near the horizon, and hence these shadows are most noticeable in that quarter of the heavens which is below the sun. This effect is of common occurrence in summer, and is known by the name of "the sun's drawing water." Occasionally we notice these shadows diverging in every direction from the sun, not only downward, but also laterally and even upward. These shadows are parallel bands, and the apparent divergence is the effect of perspective.

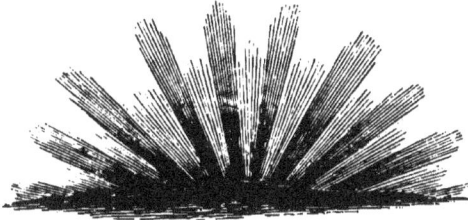

201. *Shadows after Sunset.* — A similar phenomenon is frequently noticed about fifteen minutes after sunset, when the shadows of clouds near the horizon are projected upon the western sky in the form of radiant beams diverging from the sun. These beams are parallel lines of indefinite length, but from the effect of perspective they seem to diverge from the sun, and if they could be traced entirely across the sky they would, for the same reason, converge to a point directly opposite to the sun. Such cases are sometimes, though not very frequently noticed. Similar shadows are sometimes seen in the morning before sunrise, and form a conspicuous feature of the morning twilight. This effect is sometimes noticed in nearly every part of the world. It must have attracted the attention of the ancient Greeks, and is thought to explain that poetic expression, "the rosy fingered morn."

SECTION V.

RAIN.

202. *Origin of Rain.*—When a portion of the vapor which exists in the air is condensed, a mist or cloud is formed. Generally
this condensation proceeds slowly, and the clouds which result
do not furnish rain. But when this condensation takes place
with sufficient rapidity, the small particles of mist increase in diameter by the condensation of more vapor, and, forming drops
of considerable size, they descend to the earth in a shower of
rain.

203. *Diameter and Velocity of Drops of Rain.*—Drops of rain
vary in diameter from a quarter of an inch to $\frac{1}{25}$th and even $\frac{1}{75}$th
of an inch. The velocity which they acquire in their descent is
very small. A drop falling in a vacuum would be continually
accelerated, and at the end of one minute would have the velocity of a cannon ball; but, falling through the atmosphere, the resistance increases with the velocity until this resistance becomes
equal to the weight of the drop. When this result takes place
there can be no farther increase of velocity, and the drop afterward descends with a uniform motion. A drop of rain $\frac{1}{4}$th of an
inch in diameter, by falling through the atmosphere, can not acquire a velocity exceeding 34 feet per second; a drop $\frac{1}{25}$th of an
inch in diameter can only acquire a velocity of 13 feet per second;
a drop $\frac{1}{75}$th of an inch in diameter, a velocity of 8 feet per second;
and a globule of water $\frac{1}{1000}$th of an inch in diameter can not acquire a velocity so great as two inches per second.

204. *To Measure the Amount of Rain.*—The amount of rain which
falls from the sky is measured by a *pluviameter*, or rain-gauge. The
object of the rain-gauge is to determine the average depth of rain
which falls in a given neighborhood. For this purpose we catch
in a vessel the rain which falls upon a limited space, as a square
foot, and hence infer the amount which falls in the neighborhood. It is, then, essential to the accuracy of our conclusion that
we catch *all* the rain which falls within the prescribed limits, and
no more; and also that this amount be equal to the average

Fig. 49.

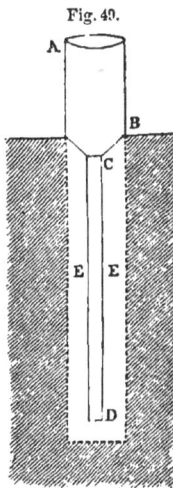

depth which falls in the vicinity. To secure the first object, the edge of the vessel should be sharp and its sides upright. If the edge of the vessel be thick, or the sides be much inclined, the rain which falls upon the edge, or upon the sloping sides, will be scattered in various directions, and a part will be wasted. A cylinder several inches deep is the most convenient form of gauge. This cylinder may be large or small. They have generally been made about ten inches in diameter; but a cylinder two inches in diameter, if carefully made, may yield very accurate results. Fig. 49 shows the gauge employed by the Smithsonian Institution. The cylinder AB is two inches in diameter, and the tube CD is about half an inch in diameter.

205. *Amount of Rain determined.* —The amount of rain collected in the gauge may be measured in a tube properly graduated by comparing the area of a section of the gauge with that of the tube. Suppose the gauge to be a cylinder ten inches in diameter. Take a glass tube exactly one inch in diameter, and graduate its side to inches and tenths, and measure the rain in this tube. One inch of water in the tube will correspond to one hundredth of an inch in the gauge, and a tenth of an inch in the tube to one thousandth in the gauge. We may thus easily measure the depth of the fallen rain to the accuracy of one thousandth of an inch. In a similar manner we may measure the depth of the rain, whatever be the diameter or form of the gauge.

206. *Proper Exposure of the Gauge.* —In order that the amount of rain collected in the gauge may be equal to the average depth which falls in the vicinity, a proper exposure is indispensable, and this is sometimes difficult to be attained. If the gauge be erected near a building it is liable to be affected by eddies or currents of air, causing more rain to fall on one side of the building than on the other. The most suitable place for a rain-gauge is in an open field remote from all obstructions; or, if it must be near a building, a position should be selected which is least exposed to the influence of eddies.

207. *Influence of Height of the Gauge.*—Two gauges placed near each other, at different elevations, do not generally collect the same quantity of rain, the lower gauge usually showing the most water. At the Observatory of Greenwich, a gauge at the surface of the ground annually collects two thirds more rain than a gauge elevated fifty feet above it. Similar differences, but less in amount, have been observed at other places in England, as well as in Paris and Philadelphia.

This result has been ascribed to an increase in the size of the drops as they descend through a humid atmosphere, the drops being generally colder than the surrounding air. But so rapid an increase in the size of a drop, amounting to two thirds in a fall of fifty feet, is altogether incredible. Moreover, it ought sometimes to happen that the drops should diminish in size by evaporation in traversing a warmer stratum of air, while observation always indicates the greatest amount of rain in the lower gauge.

This difference is probably caused by eddies formed in the air about the gauge. A portion of the air which strikes against the gauge glances up over it and spreads out laterally, carrying along with it the descending drops of rain, thus dispersing the drops which would otherwise fall into the gauge, and diminishing the quantity of water which it collects. These eddies are strongest where the velocity of the wind is the greatest; that is, they produce the greatest effect at a considerable elevation above the ground, where the course of the wind is unobstructed by opposing buildings.

Hence we conclude that the best location for a rain-gauge is to bury it in the earth, making its top just even with the surface of the ground.

208. *How Rain is Caused.*—Rain is but the condensed vapor of the air, and this condensation can only be caused by cooling the air below the temperature of the dew-point. This reduction of temperature may be effected by radiation, or by the contact of warm air with the cold surface of the ground, especially the surface of an elevated mountain; or by the mingling of warm air with colder air; but these processes are so gradual, or limited in extent, that they probably never result in any thing more than a fog or a cloud. In order to produce an abundant rain, the air must be *suddenly* cooled below the dew-point, and there is no mode in which this can be so readily accomplished as by forcing it up to an elevation of one or two miles above the earth's surface. The temperature of the air sinks about thirty-five degrees in two miles of elevation; and if air from the earth's surface should be forced up to this height, a large portion of the vapor which is carried up with the air must be condensed. Such an effect may result from an interposed mountain, or from the opposition of two currents of air. Examples of both of these methods are of daily occurrence.

209. *Hutton's Theory of Rain.*—In 1784, Dr. Hutton, of Edinburg, proposed a theory of rain, which has acquired great celebrity. This theory is founded upon the following principle: When two masses of air of different temperatures, and both saturated with vapor, are mingled together, the temperature of the mixture is too low to contain all the vapor of the combined masses. This excess of moisture must therefore be discharged in the form of rain.

Suppose, for example, there are two masses of air having the temperatures of 60° and 80°, and that each is saturated with moisture. The elastic force of vapor at these temperatures is 0.518 and 1.023; the mean of the two being 0.770 inch. Suppose the mixture to have a temperature of 70°, at which temperature the elastic force of vapor is 0.733 inch. The difference is 0.037 inch of mercury, or 0.503 inch of water, and this, it is claimed, is the amount of water that should be precipitated the moment these two masses of air are perfectly mingled. A similar result should take place if the two masses of air contain considerable moisture, but without being saturated.

It is objected to this theory that it is impossible to mingle together two large masses of air of different temperatures, except very slowly, and hence the resulting precipitation can not be considerable. Moreover, the latent heat evolved in the condensation of the vapor would raise the temperature of the mixture, so that a less quantity of water than that above supposed would be precipitated. Such a mixture might, therefore, give rise to a cloud, but never to a copious shower.

210. *Distribution of Rain over the Earth's Surface.*—The fall of

rain is very unequally distributed over the earth's surface, varying from zero to a depth of fifty feet in a year.   The amount of
rain is affected by the latitude of the place; by its elevation above

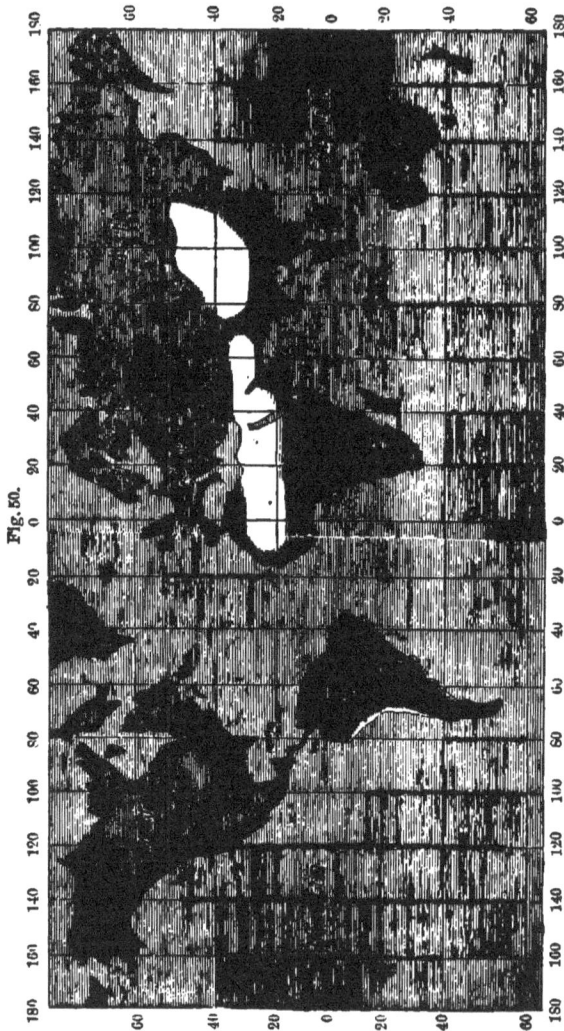

the sea; by the proximity and course of chains of mountains ¡
the proximity and configuration of the coast, as well as by the
direction of the prevalent winds.   Fig. 50 is designed to show the

distribution of rain over the earth's surface; a deep shade indicating a great fall of rain, and a light shade indicating a scarcity
of rain.

211. *Influence of Latitude.*—The average fall of rain is greatest
at the equator, and diminishes as we proceed toward the poles.
The following table shows the average annual fall of rain for
every ten degrees of latitude as far as 60°.

At the equator, 104 inches.		In latitude 40°, 30 inches.	
In latitude 10°,	85 "	" " 50°, 25 "	
" " 20°,	70 "	" " 60°, 20 "	
" " 30°,	40 "		

That more rain should fall at the equator than in high latitudes might be expected from the greater amount of vapor contained in the air. The average amount of vapor present in the
air is about five times as great at the equator as in latitude 60°;
and an equal reduction of temperature must precipitate more
moisture from the air in a warm than in a cold climate. If the
causes which produce rain acted with equal intensity in all latitudes, we might expect that the average amount of rain in each
latitude would be proportional to the quantity of vapor contained in the atmosphere. In this case, if we assume the mean
fall of rain at the equator to be 104 inches, as has been determined by observation, the annual fall in other latitudes would be
as follows:

In latitude 10°, 101 inches.		In latitude 40°, 45 inches.	
" " 20°, 90 "		" " 50°, 27 "	
" " 30°, 70 "		" " 60°, 18 "	

We thus see that while in latitude 60° the actual fall of rain is
fully equal to what might be expected from the amount of vapor
present in the air, there is a great deficiency of rain in the intermediate latitudes, which is most decided from latitude 10° to° 30;
in other words, we must conclude that the causes which produce
rain act with less intensity near latitude 30° than they do near
the equator, or latitude 60°.

212. *Number of Rainy Days.*—We shall arrive at the same conclusion if we compare the number of rainy days in a year in different latitudes. The following table, deduced from a comparison
of the log-books of a large number of vessels navigating the

II

Atlantic Ocean, shows the average number of rains which occur during a hundred days on different parallels over the Atlantic:

Between	Between
latitude 0° and 10°, 45 rains.	latitude 30° and 40°, 25 rains.
" 10° " 20°, 18 "	" 40° " 50°, 34 "
" 20° " 30°, 21 " ·	" 50° " 60°, 40 "

We thus see that near latitude 60° the number of rainy days is about the same as at the equator, and is about double what it is from latitude 10° to 30°. This result accords with that deduced in the preceding article, viz., that the causes which produce rain act with diminished intensity between the parallels of 10° and 30°.

The increased fall of rain near the equator is ascribed to the ascent of a vast column of air due to the meeting of the northern and southern trade winds, and there is a similar meeting of opposing winds near the parallel of 60°. On the other hand, between the parallels of 10° and 30°, the winds are more uniform in their direction than in any other part of the world.

The frequent rains near the equator and the parallel of 60° explain the diminished height of the barometer in those localities, for we find the barometer always stands lowest near the centre of a great rain-storm.

213. *Influence of Elevation above the Sea.*—The annual fall of rain is uniformly greater on mountains of moderate elevation than it is at the level of the sea; and at a certain height the fall is from two to three times as great as it is near the base of the mountain. On the island of Guadeloupe, in latitude 16°, near the summit of a mountain of 5000 feet elevation, the fall of rain in 1828 was 292 inches, while near the base of the mountain the fall was only 127 inches. This difference is not due to the coldness of the mountain. An equal and probably a greater effect would be produced by a volcanic mountain whose surface was covered with melted lava. When a current of air meets an interposed mountain, it is forced up the side of the mountain; that is, it is elevated above the earth's surface into a colder region, and its vapor is precipitated by the cold of elevation.

We find the same principle exemplified wherever there are high mountains. Along the western coast of Hindostan runs a range of mountains whose summits are deluged with rain, while

near their western base the amount of rain is by no means extraordinary, and on their eastern side the fall is less than one third
of the average for the same latitude. · At Bombay, on the western
side of the mountain,
the average annual fall
of rain is 78 inches; at
the elevation of 4500
feet the average fall is
254 inches, and in 1842

Fig. 51.

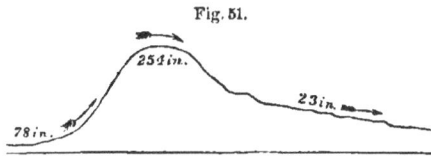

the fall amounted to 305 inches; while at Poonah, on the eastern
side of the mountain, the average fall is only 23 inches. This
rain falls almost wholly from June to October, during the prevalence of the southwest monsoon. The warm and moist air from
the ocean, encountering this range of mountains, is elevated high
above the surface of the sea, by which means it is cooled, and its
vapor is condensed over the summit of the mountain. When
this air descends on the eastern side of the mountain it is a *dry*
air, and has but little vapor remaining to be precipitated.

A similar effect takes place on the southern slope of the Himalaya Mountains, about 300 miles north of Calcutta, where, at an
elevation of 4500 feet, the fall of rain in 1851 was 610 inches, all
of which fell from April to October, during the prevalence of the
southwest monsoon.

Similar effects take place in Central America, and on several
of the West India Islands, where the prevalent winds come from
a warm sea, and contain a large amount of vapor.

214. *Maximum Fall of Rain.*—The increased fall of rain upon
mountains attains its maximum at a certain elevation, and above
that point the fall of rain decreases as we ascend. The elevation
at which the fall of rain is greatest is not every where the same.
In India it is about 4500 feet, while in Great Britain it is about
1900 feet.

215. *Influence of Proximity to a Mountain.*—Sometimes the mere
proximity to a mountain causes more rain to fall at the level of
the sea than is usually found in the same latitude. Thus, at Vera
Cruz, 278 inches of rain have been known to fall in a single year;
and the mean annual fall is 185 inches, which is fully double the
average amount for the Gulf of Mexico. This result is to be as-

cribed to the high mountains on the west coast of Vera Cruz, by which the warm and moist air from the Gulf is forced up to a great height, and its vapor is condensed by the cold of elevation, and this influence is not confined to the immediate vicinity of the mountain, but extends to some distance beyond its base.

So, also, on the Northwest Coast of America, near latitude 60°, for a similar reason, the annual fall of rain is 90 inches, which is at least four times the average for other parts of the globe in the same latitude.

For a like reason, on the coast of Norway, in latitude 60°, the annual fall of rain is more than 80 inches.

216. *Influence of Proximity to the Ocean.*—An increase of rain usually results from mere proximity to the ocean, even where there are no mountains, especially if the prevalent winds come from the sea. This effect is most noticeable near the coast, and goes on diminishing as we proceed toward the interior of a continent. Thus, in Europe, near the Atlantic coast, the fall of rain varies from 30 to 40 inches; in Central Europe it seldom exceeds 20 inches; while throughout a large part of Russia it is only 15 inches, and in Northern Asia it is still less.

Similar results, but somewhat more complicated, are found in the United States. On the Atlantic coast, near the parallel of 45°, the annual fall of rain is 40 inches; in Michigan it is about 30 inches; in Minnesota 25 inches; and near the Missouri River, on the same parallel, it is only 15 inches.

217. *Influence of Winds.*— Along the Atlantic coast of the United States, rain occurs most frequently with the wind from the northeast. Out of one hundred cases of rain or snow recorded at New Haven, the average number occurring with the different winds is as follows:

N.	N.E.	E.	S.E.	S.	S.W.	W.	N.W.
8	37	6	19	7	15	1	7

Storms at New Haven generally begin with an easterly wind and end with a westerly wind, so that the same storm is attended by both winds; but as the rain or snow with the first wind generally continues longest, the easterly wind is recorded as accompanying rain at a greater number of the regular hours of observation.

Throughout most of the interior of the United States, the principal part of the rain comes with a westerly wind. At Cincinnati, out of one hundred cases of rain or snow, the average number occurring with the different winds is as follows:

N.	N.E.	E.	S.E.	S.	S.W.	W.	N.W.
2	10	1	9	10	25	18	25

In Central Europe about three fourths of all the rain occurs with a westerly wind.

218. *Annual Fall of Rain at different Places.* — To obtain the mean fall of rain at any place requires observations continued for a considerable number of years, for it not unfrequently happens that the rain of one year is double that of some other year at the same place. The following table shows approximately the average annual fall of rain for different parts of the United States:

	Inches.		Inches.
Alabama and Louisiana .	56	Ohio	40
Oregon	49	New England	40
Florida	48	New York	37
Virginia and the Carolinas	48	Michigan and Wisconsin	32
Tennessee and Kentucky	48	Iowa and Kansas . . .	31
Georgia	44	Texas	29
Arkansas and Missouri .	42	California	18
Maryland and Pennsylvania	41	New Mexico	13

219. *Distribution of Rain throughout the Year.* — Throughout most of the United States east of the Rocky Mountains, the rain is pretty equally distributed through the different months of the year, but the rain of summer is every where somewhat greater than the rain of winter, including the melted snow. In New England the difference between the rain for these two seasons is less than 10 per cent. ; in the State of New York it is nearly 50 per cent.; in Virginia and the Carolinas it is 100 per cent.; in Florida it is 200 per cent.; in Texas it is 75 per cent.; in Ohio it is 25 per cent.; in Michigan and Wisconsin it is 140 per cent.; while in Iowa and Kansas it is 300 per cent; that is, the fall of rain in summer is four times as great as it is in winter. On the Pacific coast this law is reversed. In California the rain of winter is more than twenty times as great as that of summer, and in Oregon it is seven times as great. See Table XXIX.

220. *Rainy Season and Dry Season.*—When the rain is very unequally distributed through the different months, the year is naturally divided into the rainy season and the dry season. Throughout most of California but little rain falls except during the six colder months, and during the four months from June to September rain is almost unknown. No rain falls during the summer months, when the wind blows almost uninterruptedly from the southwest, because this air comes from a colder ocean, and, passing over the heated land, its vapor is not condensed until it meets the Nevada Mountains, on the eastern margin of California.

Wherever the direction of the prevalent wind changes greatly with the season of the year, we generally find the rain unequally distributed through the different months. On the *west* coast of Hindostan, nearly all the rain falls from April to September, during the prevalence of the southwest monsoon; but during the other half of the year, the winds coming from the northeast have already passed over high mountains, where they have lost their moisture, and descend to the earth as dry winds, which often furnish no rain for months in succession.

On the *east* coast of Hindostan, almost no rain falls during the prevalence of the southwest monsoon, but abundant rains occur during the prevalence of the northeast monsoon, when the warm air from the Bay of Bengal has a higher temperature than the land.

A similar inequality occurs at many places in tropical America. At Vera Cruz almost the entire fall of rain occurs from May to October, when the winds are easterly; but during the rest of the year the winds are northwesterly, and several months will sometimes pass without a drop of rain.

At some places near the equator there are *two* rainy periods of the year, the maxima occurring in June and December.

221. *Greatest Fall of Rain.*—There are certain portions of the globe which are habitually, and others occasionally deluged with rain. On the southern slope of the Himalaya Mountains, at the height of 4500 feet, in latitude 25°, there have been registered in a single year 610 inches of rain; and of this, 147 inches fell in the month of June. At a station in latitude 18°, near the western coast of Hindostan, the average fall for fifteen years has been

254 inches. In the northwestern part of England, at the height of 1300 feet, the average annual fall of rain is 146 inches, while at London the annual fall is only 20 inches. At Vera Cruz the annual fall is 183 inches, and 60 inches have been recorded in a single month. See Table XXXI.

222. *Remarkable Showers.*—Throughout most of the United States the rain which falls in one day rarely amounts to one inch, but occasionally the fall is much more remarkable. Thus, at Flatbush, Long Island, on the 22d of August, 1843, nine inches of rain fell in eight hours; at Catskill, New York, on the 26th of July, 1819, fifteen inches of rain fell in six hours; at Wilmington, Delaware, on the 29th of July, 1834, five inches of rain fell in two and a half hours; and at Fairfield, Ohio, on the 12th of August, 1861, eight inches of rain fell in eleven hours.

In India fifteen inches of rain have fallen in a single day, while at several places in the vicinity of Switzerland thirty inches of rain have been reported to fall in a single day.

It is not supposed that in any of these cases the amount of rain was measured with absolute precision; but that the fall was very unusual was evident from the aspect of the country after the storm.

Rains so remarkable are necessarily quite limited in extent, for, if every particle of moisture in the atmosphere were precipitated, it would cover the entire globe to a depth of less than four inches. This result is obtained as follows: The average temperature of the entire surface of the globe is estimated at 58°, and the average dew-point at 51°. At this temperature vapor will sustain a column of mercury 0.374 inch in height. The weights of equal volumes of aqueous vapor and air at the same temperature and pressure are as 5 to 8 nearly, and the specific gravity of mercury is 13.6. Hence vapor at 51°, reduced to water, becomes $0.374 \times 13.6 \times 0.624$, which equals 3.17 inches.

At the close of a long rain-storm it is not uncommon for the air to contain more moisture than it did at its commencement. Hence we must conclude that the rain which falls in these remarkable showers is derived from moist air drawn from remote places.

223. *Deserts.*—There are large portions of the earth's surface where rain is almost entirely unknown, viz., the interior of Africa between the parallels of 20° and 30°, including most of Egypt; also a considerable portion of Arabia and Persia; the great desert of Gobi, on the northeast side of the Himalaya Mountains, with portions of Peru and California.

There are also other districts where the amount of rain does not exceed one tenth of that which is found elsewhere in the same latitude, such as Lower California, where the annual fall of rain is only three inches; also the northern coast of Africa, Lower Egypt, and Persia. See Table XXX.

224. Cause of the African Desert.—The Great Desert of Africa lies near the northern limit of the trade winds, where, as we have already seen, the causes which produce rain act with the least energy. This desert is an immense sandy plain, with a range of mountains near its northern, as well as its southern border. When the N.E. trade wind first strikes the continent of Africa, a portion of the vapor is condensed on the northern mountains. As the wind proceeds southward, it advances toward a warmer latitude, which has a greater capacity for moisture; and there are no mountains or opposing winds to force the air up above the earth's surface until we approach the parallel of 10°, where we find a long chain of mountains, over which the vapor is condensed in copious rains. The heat which is liberated in the condensation of this vapor is one cause of the steady trade winds, and the absence of rain over the Desert. Here and there in the midst of the Desert is found a high peak, or small mountain, and here rain is occasionally seen.

Similar considerations explain the small amount of rain in Egypt and Arabia.

225. Great Desert of Gobi, etc. — The great desert of Gobi is caused by the Himalaya Mountains. Here the prevalent winds are from the S.W., and, having just passed over the mountains, they have lost nearly all their vapor, that is, they are extremely dry winds, having little moisture to be precipitated.

Peru is situated within the region of the S.E. trade winds, which, on meeting the Andes, are forced up to such an elevation that their moisture is nearly all condensed, and they descend on the Pacific side as dry winds, and have no moisture which can be condensed at the temperature which prevails in Peru. The principal tributaries of the Amazon are fed by the rains which fall on the windward side of the Andes.

Between the two great mountain ranges, the Sierra Nevadas and the Rocky Mountains, comprehending portions of Utah, New Mexico, and California, is a region which is almost entirely destitute of rain. Throughout this region, whether the wind blows from the east or the west, it has lost most of its vapor by passing over the mountains. It is therefore a dry air, and has but little vapor to be precipitated.

So, also, on the east side of the Rocky Mountains, the prevalent winds, being westerly, have lost their vapor by passing over the mountains, and the country is a barren desert, almost without rain.

226. *Rain without Clouds.*—Ordinarily clouds seem to be the reservoirs from which the rain descends, but rain has been known to fall when no cloud could be seen near the zenith, and even at times when no cloud appeared above the horizon. Thus, on the 23d of April, 1800, at 9 P.M., rain fell for twenty minutes at Philadelphia, although the heavens immediately overhead appeared perfectly clear, and the stars shone with undiminished lustre. Not a cloud could be seen within 15° of the zenith. Also on the 9th of August, 1837, a shower fell at Geneva, Switzerland, which lasted two or three minutes, although the sky was cloudless. Many similar cases have been observed in other parts of the world.

227. *Rain from Clouds not in the Zenith.*—It is probable that in some cases rain reaches the earth's surface from clouds removed several degrees from the zenith. The path of a rain-drop often makes an angle with the vertical greater even than 45°, and rain might therefore reach the earth from a cloud removed 20° or 30°, and perhaps even farther from the zenith, especially if there prevailed near the earth's surface a fresh breeze, blowing in a direction different from that of the current which conveys the cloud. This principle will probably explain some of the cases which have been reported; but there are other cases in which it is said that rain has fallen, although no cloud was visible above the horizon.

228. *Rain from Translucent Clouds.*—It is probable that, in

these cases of remarkable rain-falls, although the sky was free
from dense clouds, such as entirely conceal the stars, it was not
entirely free from a haziness, which is, indeed, nothing else than
a cloud so thin as to allow the brighter stars to shine through it.
The partial transparency of such a cloud may be due to the small
number and large size of the rain-drops.

Pure water is nearly transparent, and a fog is opaque, simply
on account of the minuteness and consequent multitude of the
condensed particles. A certain amount of light is reflected from
the surface of each particle, and in a fog the number of reflecting
surfaces is so great that a beam of light is wholly reflected before
it can penetrate through the mass. But if the amount of water
which composes a fog were all collected in a few large drops, the
number of reflecting surfaces would be comparatively small; that
is, they would but slightly affect the transparency of the air. It
is probable, therefore, that when rain falls from a cloudless sky,
the vapor is condensed in a few large drops, instead of a multitude of minute ones. This condensation probably takes place
with great suddenness in the lower strata of the air, which was
previously saturated with moisture.

229. *Snow from a Cloudless Sky.*—In the polar regions a fine
snow sometimes falls from a cloudless sky. So, also, in New England, during a period of intense cold, we sometimes see flakes of
snow descending from the sky, although there is no cloud sufficient to obscure the sun or moon, or even the light of the brighter stars. In such cases, the vapor rising from the earth is probably condensed before it attains a great elevation, and both the
thickness and density of the cloud are quite small.

SECTION VI.

SNOW.

230. *How Flakes of Snow are formed.*—When the vapor of the
air is precipitated at a very low temperature, the vapor is condensed in the solid state, without passing through the condition
of a liquid, and generally assumes the crystalline form. These
minute crystals of ice attach themselves to each other and form
flakes of snow, which descend very slowly to the earth. When
the lower stratum of the air is much above 32°, the flakes of snow

melt before they reach the ground, so that rain may frequently be seen falling on an open plain, while from the same cloud snow is falling upon a neighboring mountain.

During the severe cold of winter we may frequently witness snow produced artificially. When a large number of people are assembled in the same hall, and the room being uncomfortably warm, a window is opened, the warm air of the room is frequently condensed by the cold external air, and falls to the ground in the form of flakes of snow of extreme delicacy.

231. *Where Snow Falls.*—Within the torrid zone snow is almost never seen, except on elevated mountains, because near the level of the sea the temperature is above the freezing point. For a similar reason, in the middle latitudes, the fall of snow occurs only in winter, while in the polar regions nearly all the moisture which is precipitated descends to the earth in the form of snow.

The zone within which snow never falls is determined not so much by the mean temperature of the year, or the mean temperature of the coldest month, as by the temperature of the coldest day of winter. At all places where the thermometer in winter sinks much below 32°, snow may occasionally fall. The boundary of the zone within which snow does not fall, except in a few very rare cases, is an undulating line crossing the Pacific coast of America near lat. 39°, and the Atlantic coast near lat. 35°; it passes near Gibraltar in lat. 36°, and on the coast of China descends to lat. 24°, which is but a little north of Canton.

A slight fall of snow occasionally occurs at San Francisco, California; it occasionally falls at New Orleans, and also at Galveston, lat. 29°; and snow sufficient for sleighing has been known at Charleston, S. C. Snow has also been known to fall at Canton, within the torrid zone, to the depth of four inches.

232. *Annual Amount of Snow.*—The amount of snow which falls in a year varies in different localities from zero to twelve feet. In Spitzbergen the annual fall of snow is from three to five feet. In the State of Maine the average annual fall of snow is seven and a half feet, and the amount in a single year has been known to exceed twelve feet; but this amount is not all seen at the same time. In Vermont and New Hampshire the annual fall is six feet. In Central Massachusetts the annual fall is four and a half feet, and

the snow has been known to lie five feet on a level.   In Connecticut the average fall is three and a half feet; in New Jersey, two
and a half feet; in Southern Ohio, one foot and a half; and in
Iowa, one foot.

Snow recently fallen has a very small specific gravity, for a
foot of snow, when melted, furnishes only one inch of water.

233. *Form of Snow-flakes.*—Crystals of ice generally exhibit
the form of long needles or spiculæ, each being a slender prism
with angles of 120°.   These crystals are often seen in great perfection in hoar-frost.   Flakes of snow generally consist of combi-
nations of spiculæ and of thin plates or laminæ of ice, which usually present angles of 60° or 120°.   Sometimes we find simply six
spiculæ combined in angles of 60°, forming a star with six rays.
Sometimes to each of these spiculæ are attached shorter spiculæ,
also inclined at angles of 60°, in number amounting to 2, 4, 6, etc.,
up to a dozen or more, forming a perfectly symmetrical figure
bearing some resemblance to a flower of great complexity.   See
the first six forms in Fig. 52.

Fig. 52.

Sometimes we find a simple lamina of ice, in which case the form is usually that of a regular hexagon, which sometimes has the appearance of being composed of equilateral triangles. Sometimes ice spiculæ are attached to the angles of the hexagon ; sometimes attached to the angles of a central hexagon we find six smaller hexagons, or perhaps rhomboids composed of two equilateral triangles. Sometimes the central figure consists of six such rhomboids, with ice spiculæ or other rhomboids attached to the angles.

Sometimes the flakes present forms which can not apparently be resolved into any of the preceding elements. ,.

Several hundred different forms of snow crystals have been observed and figured. Fig. 52 presents a specimen of the simplest forms, and also of the most complicated. These crystals are seen in their greatest perfection when the air is tranquil, cold, and dry.

234. *Size of Snow-flakes.*—Snow-flakes vary in size, according to the temperature at which they are formed. If formed at a very low temperature, their diameter is often less than one tenth of an inch ; when formed near the temperature of 32°, they are sometimes found one inch in diameter.

235. *Natural Snow-balls.*—Sometimes a vast number of snow-flakes attach themselves together, and descend to the earth as a loose snow-ball one or two inches in diameter. Sometimes, after the snow has fallen, it is driven along by the wind, and is rolled into balls of vast size. These balls are usually cylindrical, somewhat hollowed in the centre, and they have been known to attain a diameter of three feet. They are of common occurrence on the slopes of the Alps, in Switzerland.

236. *Snow White and Phosphorescent.*—Since snow is but frozen water, it might be expected that it would be transparent like water, or large blocks of pure ice. Its brilliant whiteness is due mainly to the number of reflecting surfaces arising from the small size of the spiculæ of ice. In the same manner, the most transparent glass loses its transparency when pulverized.

Snow is feebly *phosphorescent.* This is proved by the fact that in the darkest nights, when the ground is covered with snow, the snow appears more luminous than the sky. Its light can not, therefore, be simply the reflected light of the sky. This phosphorescence appears to be in part acquired by exposure to the rays of the sun during the preceding day. If, on the morning of a clear day, we cover a portion of the snow with an opaque screen, and uncover it at evening, we shall find that this portion is somewhat less luminous than the surrounding snow. Snow, like many other substances, after being exposed to a bright light, retains a portion of the light for some time after the source of light is withdrawn.

237. *Red Snow in the Polar Regions.*—In those places where snow lies unmelted from one year to another, it sometimes acquires a ruddy color, and occasionally becomes *red* like blood. This occurs in the polar regions, and also on the mountains of Southern Europe. In Spitzbergen the snow sometimes appears of a green hue. It has been discovered that these colors are due to a vegetable production resembling a mushroom, which is excessively minute, not exceeding $\frac{1}{1000}$th inch in diameter. There is, then, a species of vegetation which may flourish at a temperature which never exceeds that of melting ice.

238. *Glaciers.*—The summits of high mountains, even under the equator, are covered with perpetual snow. Within the tropics the limit of perpetual snow varies from 16,000 to 18,000 feet, while on the Alps of Switzerland it varies from 8000 to 9000 feet. On these mountains the snow accumulates from year to year, and in sheltered ravines, where it can not be blown away by the wind, acquires an immense thickness. Under continued pressure this snow becomes solidified, so as to acquire the density of compact ice. The gorges of the Alps are filled with ice of this description, which is known by the name of *glaciers*. These glaciers are from five to ten or more miles in length, and they follow the gorges from the summit of Mount Blanc down to the base of the mountain. They are frequently half a mile or more in breadth, and have a thickness of 200 to 5000 feet. This ice, sustaining the pressure of a long column, rising to the height of 10,000 or 12,000 feet, is crowded down into the valleys, so that the entire glacier has a descending motion like a river. The principal glacier of Switzerland has a descending motion which in some places amounts to 876 feet in a year, and in other places only 274 feet.

This motion is continuous, and is probably never wholly interrupted. Nevertheless, the motion is greatest in summer and least in winter, and the velocity increases with the angle of descent. The middle of the glacier generally moves faster than the sides. These glaciers extend down into the valleys, where the temperature is such as to allow wheat and potatoes to come to maturity, and a traveler may sometimes stand upon the edge of a glacier and pick ripe cherries from a tree. The ice melts, indeed, under a summer's sun, but the waste of summer is supplied by the slow motion of the descending mass, so that the lower end of the glacier remains nearly stationary from age to age. Fig. 53 represents one of the most remarkable glaciers of the Alps. It is seen to be intersected by numerous fissures, caused by its motion down an irregular valley.

Fig. 53.

The total number of glaciers among the Alps is estimated at between 500 and 600, and they cover an area of nearly 1500 square miles. The lowest of the glaciers of the Alps descends to the level of 3400 feet above the sea.

239. *Geographical Distribution of Glaciers.*—No glaciers have been found within the tropics, but they are common on the high mountains of the middle latitudes, and especially in the polar regions. The glaciers of the Himalayas are very numerous and of immense extent, and are the sources of large rivers. In lat. 27° they descend to the level of 13,000 feet, and in lat. 36° they descend to the level of 9000 feet.

The Pyrenees are nearly destitute of true glaciers.

The elevated mountains of Greenland are covered with perpetual snow and ice, which in many places extends to the sea-shore. The snow of winter becomes solidified by the warmth of summer, acquiring in time the density of ice. This ice is crowded down by its own weight into the sea, and sometimes extends several miles beyond the original shore-line. By the buoyant power of the water the outer end of the glacier is lifted, and after a time a mass, perhaps a mile or more in diameter, is cracked off. This mass is drifted southward to the middle latitudes, and is called an iceberg. An iceberg has been measured three fourths of a mile square, and 315 feet high. Large icebergs continue unmelted for many weeks, and sometimes advance to lat. 36°.

In Norway the glaciers are numerous, and near lat. 60° one of them descends to within 150 feet of the sea level, while in lat. 70° they descend into the sea.

In Spitzbergen one glacier presents a front of eleven miles to the sea, with a cliff 400 feet high, and extends backward to the mountains.

The interior of Iceland is covered with glaciers.

On the west coast of Patagonia glaciers are numerous, and in lat. 46° S. they descend to the sea.

The glaciers of Victoria Land, lat. 70 to 79° S., are even more extensive than those of Greenland.

240. Avalanches of Snow.—The snow which during winter accumulates on the sides of the Alps and other mountains, becomes softened during the summer, and frequently descends into the valleys in large masses called *avalanches*. During summer these avalanches are of hourly occurrence on some parts of the Alps, sweeping down a slope of several miles into the valleys, and are among the chief dangers encountered by travelers who attempt to climb the mountains.

SECTION VII.

HAIL.

241. *Sleet.*—In the middle latitudes, in the cold months of the year, during gusty weather, there often fall from the sky small spheres of ice, having a diameter of one twelfth to one sixth of an inch. They are generally soft, opaque, and of a whiteness approaching that of snow. The largest are sometimes surrounded with a slight film of ice. Sometimes small hailstones consist entirely of transparent ice, and these are probably rain-drops falling from clouds brought by south winds, which freeze in traversing cold strata of air near the earth.

The small hailstones of winter are termed *sleet*, to distinguish them from large hail, which falls under different circumstances.

242. *Large Hail.*—Large hail seldom if ever falls except during thunder-storms. It falls at the commencement of the storm or during its continuance. It very rarely follows rain, especially if the rain has continued for some time. The area covered by the rain-storm is much larger than that covered by the hail, and the hail at any one place continues but a very short time, generally only five or ten minutes, seldom so long as fifteen or twenty minutes.

In the United States large hail falls chiefly in summer and the latter part of spring. In India hail falls chiefly in the four months from February to May.

Hail falls at all hours of the day and night, but large hail is most common about the hottest part of the day, that is, about 2 P.M. The fall of large hail is commonly preceded by an unusual degree of heat. An extraordinary rise of the thermometer in April or May affords reason to anticipate a hail-storm.

243. *Size of Hailstones.*—The size of hailstones varies from one tenth of an inch or less in diameter to more than four inches. On the 13th of August, 1851, about 1 P.M., hailstones fell in New Hampshire weighing 18 ounces. A sphere of solid ice weighing 18 ounces has a diameter of four inches, and a circumference of $12\frac{1}{2}$ inches. In the present case the stones were somewhat porous and of irregular shape, and their largest circumference ex-

T

ceeded 15 inches. A few years since, hailstones weighing sixteen ounces fell in the city of Pittsburg, and hailstones weighing over half a pound have fallen in several places of the United States.

On the 7th of May, 1822, there fell at Bonn, in Germany, hailstones weighing from twelve to thirteen ounces, and stones weighing half a pound have repeatedly fallen in France and Italy.

Large hail is of common occurrence in India. On the 11th of May, 1855, about 6 P.M., near the Himalaya Mountains, in latitude 29°, hailstones fell weighing from eight to ten ounces, and one or two weighed more than a pound.

On the 22d of May, 1851, in latitude 13° north, in the southern part of India, many hailstones fell about the size of oranges. The next morning, in a dry well, there was found a block of ice measuring 4½ feet long, 3 feet broad, and 18 inches thick. It is not supposed that this ice fell from the sky in a single block, but after their fall the separate hailstones became cemented together so firmly by ice as to form one solid block. Similar masses of ice derived from hail have been repeatedly seen in India, and also in the United States.

244. *Quantity of Hail.*—The quantity of hail which falls from the sky in a single shower is sometimes enormous. In the New Hampshire storm of 1851 the average depth of the hail was *four* inches. In a storm which passed over the Orkneys, on the north of Scotland, July 24th, 1818, the depth of the hail was nine inches. On the 17th of August, 1830, in the streets of Mexico, hail fell to the depth of *sixteen* inches.

245. *Form of Hailstones.*—Hailstones are ordinarily of a spheroidal form; sometimes they are oval, sometimes flattened, and sometimes of a very irregular shape. Very large hailstones often present remarkable protuberances. They often consist of an irregular assemblage of angular pieces of ice, which individually do not exceed the size of walnuts, but cemented firmly together, forming a mass as large as an orange, and sometimes as large as a turkey's egg. These small portions generally indicate a tendency to crystallization. Sometimes hailstones are studded with crystals in the form of hexagonal prisms, and when the angles melt away the prisms become nearly cylindrical. The following figure represents a hailstone which probably consisted originally

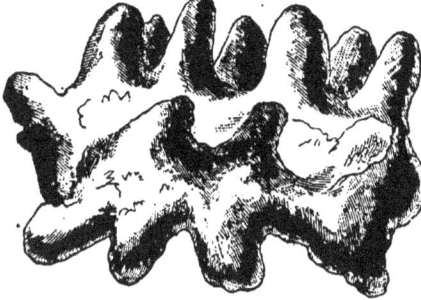

Fig. 54.

of numerous prisms cemented together, but it became so modified by melting during its fall as nearly to obliterate the crystalline structure. Sometimes hailstones have the form of pyramids, whose angles are rounded by a partial melting, and whose base is a portion of an irregular spherical surface.

246. *Structure of Hailstones.*—The centre of large hailstones usually consists of hardened snow, and this is surrounded by a coat of transparent ice. Sometimes we find alternate layers of

Fig. 55.

opaque snow and transparent ice. Often hailstones exhibit a radiated structure, resulting apparently from rows of air-bubbles disposed in radii from the centre. Sometimes large hailstones consist of very transparent and solid ice with numerous air-bubbles. Fig. 56 represents a section of a hailstone whose external appearance is represented in Fig. 55. Hailstones with a radiated structure, when broken, incline to divide into spherical pyramids, with layers parallel to their base, and this is probably the origin of pyramidal hailstones. The rupture of the spherical hailstone may be due to the sudden expansion experienced in passing from an exceedingly cold to a comparatively warm atmosphere.

Fig. 56.

247. *Geographical Distribution of Hail.*—Within the tropics hail is of rare occurrence at the level of the sea, but when it does occur the stones are generally of very great size. It becomes more common at the height of 1500 feet. In India, hail is very common on the mountains, and occurs occasionally at the level of the sea, even south of latitude 20°.

Hail is most common in the middle latitudes. In Europe hail occurs most frequently near the Atlantic coast, and diminishes in frequency as we proceed eastward. In France hail falls, on an average, fifteen times in a year; in Germany, five times; and in Russia only three times. Hail falls in every part of the United States, but cases of very large hail occur but seldom. Hail falls occasionally, but not often, in the West India Islands.

248. *Track of Hail-storms.*—Hail-storms usually travel rapidly over the country in straight bands of small breadth, but considerable length. The track of the New Hampshire storm was several miles in length, but only two miles in breadth. The track of the Orkney storm was twenty miles long and a mile and a half wide, and the storm traveled at the rate of forty miles per hour.

On the 13th of July, 1788, a hail-storm traveled from the S.W. part of France to the shores of Holland at the rate of 46 miles per hour. There were two distinct bands of hail, the breadth of that in the west being eleven miles, and that in the east six miles, with a space of fourteen miles between them. The fall of hail upon these two bands was not exactly contemporaneous, but one preceded the other about fifteen minutes. Rain fell on the outside of these bands of hail, as well as on the space between them. Each band of hail extended a distance of about 500 miles. Figure 57 represents a portion of the track of this storm in the neighborhood of Paris. The dotted bands represent the track of the hail, while the three shaded bands represent the area of the rain.

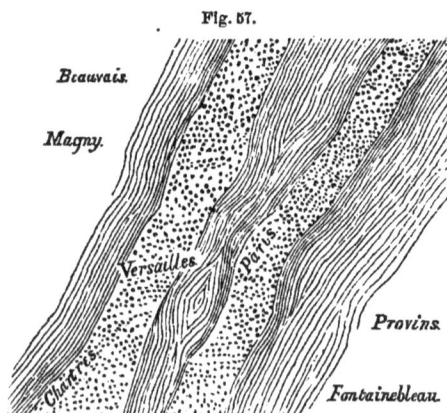

Fig. 57.

249. *Height at which Hail is formed.* — Observations made in mountainous countries have enabled us to determine nearly the elevation at which hail is formed. Small hail is of common occurrence on the summit of Mount Blanc, 15,744 feet above the level of the sea, but large hail has never been seen there. In India, at the height of 8000 feet, hailstones have fallen of sufficient size to do considerable damage.

In 1835, hailstones weighing eight ounces fell at the base of a mountain in the southern part of France, while only small hail fell at the height of 4000 to 5000 feet. From these and similar observations, it is inferred that, in the middle latitudes, hail often begins to form at an elevation exceeding 16,000 feet, but attains its greatest size below the height of 5000 feet.

250. *Origin of the Cold which causes Hail.* — The cold which congeals such large masses of ice in summer is mainly due to elevation. The temperature of hailstones at the instant of their fall has often been found below 32°, and sometimes as low as 25° F. They must, then, have been subjected to a temperature considerably below that of melting ice, probably to a temperature as low as 20° F. In the neighborhood of New York, at the height of 18,000 feet, the average summer temperature is 20°, and it is believed that during the formation of hail the temperature of the upper air is considerably below the mean.

251. *Noise preceding the Fall of Hail.*—Some seconds before the fall of hail, and occasionally several minutes, a peculiar crackling noise is often heard in the air. It has been compared to the noise of walnuts violently shaken up in a bag. This noise has been ascribed to the great velocity with which the hailstones are driven through the air, while some have ascribed it to feeble electrical discharges from one hailstone to another, for electricity always attends the progress of a hail-storm.

252. *Hail attended by Two Currents of Air.*—The formation of hail is invariably attended by two distinct currents of air, and one of these currents displaces the other with great violence. The current of air which precedes the approach of a hail-storm is extremely hot, and highly charged with moisture; that which succeeds the fall of hail has an icy chillness. The warm and humid air is displaced by the cold current, and is thus forced up to a great elevation above the earth, by which means its vapor is suddenly condensed. Upon the front of the hail-cloud this condensed vapor exists in the form of water, whose temperature is near 32°. In the interior of the hail-cloud the vapor is precipitated in the form of snow, whose temperature is sometimes as low as 20°.

253. *Process of the formation of Hail.*—Observations on the summits of mountains have shown that, on the front of the hail-cloud, there exists a violent whirling motion about a horizontal axis. This whirling motion causes the snow to collect in small balls, each of which forms the nucleus of a hailstone. The snow-ball is forced into the warm current, where it receives a layer of water, which is congealed by the cold of the nucleus, thus rendering the snowy centre more compact, and adding a shell of transparent ice. By means of the whirling motion, the hailstone, covered with a stratum of uncongealed water, is hurled into the snow-cloud, where it receives a layer of snow, and again becomes thoroughly chilled. Thence it escapes again into the water-cloud, and is covered with a layer of water, which is congealed by the cold of the nucleus. Thus, by the whirling motion, it is plunged alternately into the snow-cloud and the water-cloud, while each alternation furnishes a layer of spongy ice and a layer of transparent ice. Thus the stone grows with immense rapidity, and in a few minutes becomes a large ball, three or four inches in diameter. This oscillatory motion of the hailstones, on the front of the hail-cloud, was distinctly observed by M. Lecoq in 1835 on the summit of a mountain in the southern part of France.

254. *How Hail is sustained in the Air.*—The hailstones are sustained in the air by the violent upward motion caused by the cold current displacing the warm one. A sphere of ice two inches in diameter, by falling through a tranquil atmosphere, soon acquires a velocity of 90 feet per second. A hailstone of irregular shape would experience more resistance than a sphere, and would acquire a somewhat less velocity, but it would still fall from a height of 18,000 feet in about three minutes, which time is too small to allow the formation of masses of ice weighing one pound. An upward current of air rising with a velocity of 90 feet per second would sustain a sphere of ice two inches in diameter, and would greatly reduce the velocity of stones of larger size.

255. *How long may Hailstones be sustained?*—The strong upward movement which always attends the formation of hail is probably sufficient to sustain hailstones of the largest size as long as they can be kept within the influence of this vortex. A period of ten minutes is probably sufficient for the formation of hail-stones of the largest size. After escaping from the influence of this vortex, small stones would fall to the earth, from an elevation of 5000 feet, in about two minutes, and very large stones in one minute.

256. *Origin of the Parallel Bands of Hail.*—It is not uncommon for two or even three such vortices to form on the same day, and nearly at the same hour, at places not very remote from each other, thus forming parallel bands of hail separated by an interval of from 10 to 100 miles. Such was the French storm of 1788, and similar cases have frequently occurred in the United States.

257. *Origin of Sleet.*—The small, spongy hail of winter is probably formed in the same way as the large hail of summer; but since, in winter, the amount of vapor present in the air is small, the amount of precipitation is small, and the hailstones can never attain a large size.

258. *Hail-rods.*—It has been proposed to preserve vineyards and valuable farms from the ravages of hail by erecting an immense number of poles, armed with iron points, communicating with the earth, for the purpose of drawing off the electricity of the clouds. Multitudes of these hail-rods were formerly erected in Switzerland, but without the expected success.

It is believed that electricity performs altogether a subordinate, if not an unimportant part in the formation of hail ; and if we could draw off all the electricity from the hail-cloud as fast as it was generated, it is not improbable that hail would be formed about as large and as abundantly as at present.

But, even supposing electricity to be the sole agent in the production of hail, hail-rods could not be expected to furnish security against hail unless an entire continent could be studded thick with them, for in the middle latitudes the hail-cloud advances eastward with a velocity sometimes of 40 or more miles per hour, and the hailstones which fall in one locality are those which were forming when the cloud was many miles westward of that point ;

so that, to protect a small spot, the whole country for many miles westward should be armed with rods; and it is conceivable that a hail-cloud arriving over a region studded with these rods might immediately pour down a large quantity of hailstones which would have fallen farther eastward if the rods had not discharged the electricity of the cloud.

CHAPTER VI.

STORMS, TORNADOES, AND WATER-SPOUTS.

SECTION I.

THEORY AND LAWS OF STORMS.

259. *What is a Storm?*—Any violent and extensive commotion of the atmosphere is called a storm. Storms are usually attended by a fall of rain, or snow, or hail, and frequently by thunder and lightning; but, although it is probable that a precipitation of vapor always takes place over some portion of the area of every violent storm, yet the storm often extends beyond the area of rain or snow.

260. *Cause of Storms.*—Storms are caused by a strong and extensive upward motion of the air, by which means its vapor is condensed by the cold of elevation.

The atmosphere receives heat from the sun, and it loses heat by radiation. Only about one fourth of the rays of the sun are absorbed in passing vertically through the atmosphere. The remaining three fourths are absorbed by the earth's surface, by which means its temperature is raised, and heat is thence communicated to the air which rests upon the earth. The atmosphere thus receives its heat chiefly at the bottom, and, in consequence of radiation, loses it most rapidly at the top.

Since the density of the air is diminished by an increase of heat, the atmosphere is in a state of unstable equilibrium, and the lower strata tend continually to rise and take the place of the upper. Such ascending currents are formed on every tranquil day. As the air ascends it comes under diminished pressure and expands, and as it expands it cools at the rate of about 38 degrees for two miles of ascent. This ascending air carries with it the vapor which it contained at the earth's surface, and, if it rises high enough, the cold produced by expansion will condense a portion of this vapor into cloud. The height to which the air must ascend before it will become cold enough to form cloud depends upon the difference between the dew-point and the temperature of the air. If the dew-point be ten degrees below the temperature of the air, cloud will begin to form when the ascending current has risen about 1500 feet.

261. *Latent Heat liberated.*—As soon as a cloud begins to form, the latent heat of the vapor is liberated. To convert water into vapor requires a great amount of heat, and this heat is not appreciable to a thermometer; hence it is called latent heat. When the vapor returns to the condition of water, this heat is liberated, and becomes sensible heat; and when a cubic foot of water is precipitated from the air, as much heat is liberated as would be required to convert that amount of water into vapor. When one inch of rain falls from the sky, the amount of water precipitated exceeds two millions of cubic feet for each square mile of surface; and over each square mile of surface as much heat would be liberated as would be required to evaporate two millions of cubic feet of water.

By the heat thus liberated in the production of a cloud, the air in the cloud is warmed; it expands in volume, and the cloud continues to ascend as long as its temperature is greater than that of the surrounding air. As the cloud ascends, more vapor is condensed, while the latent heat evolved raises still farther the temperature of the air in the cloud.

262. *Shape of the Cloud thus formed.* — When, in consequence of an ascending column of air, a cloud begins to form, it is seen to swell out at the top, but its base continues at the same level; that is, the base is flat, even after the cloud has acquired great vertical height. The motion of the air here described is illustrated by Fig. 58, on the following page. During a warm and tranquil day many of these ascending columns are formed, and two or more adjacent columns often unite to form a single column.

The clouds thus formed during the day often subside and

Fig. 58.

dissolve at evening, when the surface of the earth becomes cooled by radiation, and thus a cloudy day is often followed by a cloudless evening; but when the atmosphere is unusually heated, and contains a large amount of vapor, the ascending columns generally go on increasing until rain descends.

263. *Why the Barometer falls under the Cloud.*—The expansion of the air in the forming cloud, particularly after rain begins to fall, causes the air to spread out in all directions above, causing a barometer under the middle of the cloud to fall below its mean height, and beyond the limits of the cloud to rise above its mean height. Near the limits of the cloud, the air, in consequence of its greater weight, sinks downward, and a portion of it flows along the earth's surface toward the centre of the ascending column, while beyond the limits of the cloud there is a gentle wind outward from the cloud. Since the air spreads out more rapidly above the cloud than it runs in below, storms tend to increase in diameter, and they often extend with great rapidity until they cover an area more than a thousand miles in diameter.

264. *Observations of the Barometer represented upon a Map.* — For the purpose of discovering the laws of storms, observations have been extended over a large portion of the earth's surface. In order to have a summary of all the barometric observations presented conveniently to the eye, we spread before us a map of the country, and draw upon it a line connecting all those places where the barometer at a given instant stands at its mean height. We draw another line connecting all those places where the barometer stands half an inch below the mean, another where the barometer stands half an inch above the mean, etc. Such lines show at a glance where there is an excess and where there is a deficiency of pressure, and what is the amount of this excess or deficiency.

265. *Amount of the Barometric Depression.*—Storms are often experienced simultaneously over large portions of the earth's surface. The storms of winter are particularly severe and extensive. The following remarks are restricted to winter storms, because their laws are best understood, although it is probable that winter storms do not differ materially from summer storms, except in their extent and severity.

Storms are generally accompanied by a considerable depression of the barometer below its mean height, and are succeeded by a rise of the barometer above its mean height. During the passage of a winter storm over the middle latitudes of North America, the barometer frequently sinks below its mean height over an area more than a thousand miles in diameter.

This area of low barometer is sometimes nearly circular; more frequently its form is very much elongated, its length being two or three times its breadth; and in the United States, the longer axis of this oval is uniformly turned in a north and south direction. Sometimes the barometer sinks below its mean height over an area extending 3000 miles in a north and south direction, and 1000 miles in an east and west direction. The area over which the barometer sinks *half an inch* below the mean sometimes extends 800 miles from north to south, and 400 miles from east to west. At the centre of the storm the barometer sometimes sinks *an inch* below its mean height.

Beyond the area of low barometer, the barometer rises above its mean height frequently to the amount of half an inch, sometimes to the amount of an entire inch, and occasionally still higher.

266. *Atmospheric Waves and Ocean Waves compared.*—If these inequalities of pressure were due, not to a change in the elastic force of the atmosphere, but simply to a change of height, and the atmosphere were a visible substance, then an observer sufficiently elevated above the earth might see vast depressions and elevations in the atmospheric envelope of the earth bearing some analogy to the waves of the ocean during a storm, but having vastly greater dimensions. The waves of the ocean have a breadth of a few rods and a length of a few miles, while the waves of the atmosphere sometimes have a breadth of one or two thousand miles, and a length of several thousand miles.

267. *Gradual Rise and Decline of Storms.*—Winter storms commence gradually, and generally attain their greatest violence only after a lapse of several days. After a certain period their violence gradually diminishes, and at length they disappear entirely. This succession of changes requires a period of several days, sometimes one or two weeks, and possibly even longer.

Sometimes all these changes are experienced over the same country; that is, the storm makes no progress from place to place. More commonly, however, the storm travels along the earth's surface, and although the same storm may continue for one or two weeks, or even longer, its duration at any one place may not exceed one or two days.

268. *Direction and Rate of Progress.*—Throughout the middle latitudes of this continent, when violent storms advance with considerable rapidity, the direction of progress is always from west to east. This direction is not absolutely uniform, but has been observed to vary from about due east to north 54° east.

The *rate* of progress of storms has been observed to vary from zero to forty-four miles per hour. They generally travel from St. Louis to New York in about twenty-four hours, and from New York to Newfoundland in another twenty-four hours.

Generally, when the barometer is unusually low at New York, it is unusually high at St. Louis, and also high in Newfoundland.

When a storm is about stationary, the form of the area of low barometer is nearly circular; but when the storm travels rapidly, this area is generally compressed in an east and west direction. The winter storms of the United States are therefore said to move side foremost.

269. *Fall of Rain or Snow.*—Great and sudden depressions of the barometer are almost invariably accompanied by a fall of rain or snow, and the area of greatest rain or snow corresponds nearly to the region of greatest barometric depression. Rain and snow are produced under circumstances exactly alike, with the exception of temperature; and the same storm frequently furnishes snow in the northern part of the United States, and rain in the southern part.

270. *Direction of Wind on different sides of a Storm.*—Since the tendency of the wind is always *from* a region of high barometer *toward* a region of low barometer, the wind must every where tend inward toward the centre of a violent storm; in the same manner as when a quantity of water is dipped from a tranquil lake, the surrounding water immediately flows in to restore the level surface.

But the currents of air thus set in motion toward the centre of a storm can not proceed directly toward that centre. On the north side of that centre the air is moving from a parallel which has a *less* velocity of rotation eastward toward a parallel which has a *greater* rotary velocity. It therefore has a relative motion toward the west, and that which would have been a north wind if the earth did not rotate upon its axis, becomes a northeast wind. So, also, on the south side of the centre, the air is moving from a parallel which has a greater rotary velocity toward a parallel which has a less velocity. It therefore has a relative motion toward the east, and that which would have been a south wind if the earth did not rotate upon its axis, becomes a southwest wind.

The wind, therefore, instead of blowing from every point of the compass directly toward the centre of the vortex, moves spirally inward, making a great circuit round the centre; and in the United States this rotation is in a direction contrary to that of the hands of a watch, or, as it is called, from right to left.

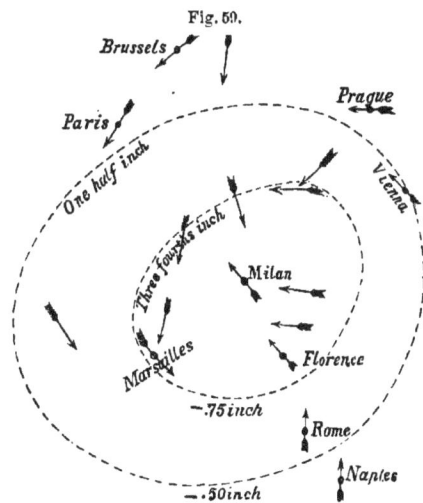

Fig. 59.

The force of the wind is generally proportional to the amount and rapidity of the depression of the barometer.

271. *A European Storm.* — The direction of the wind at the earth's surface is greatly influenced by the irregularites of the earth, as well as by local differences of temperature and moisture; nevertheless, observations of vio-

lent storms show that the prevalent direction of the wind corre- ·
sponds to the preceding principles. Figure 59, on the preceding
page, represents the winds as actually observed near the centre
of a violent storm of rain and snow which was experienced in
Central Europe on the 25th of December, 1836. The smaller oval
shows the area within which the barometer was depressed three
fourths of an inch below the mean, and the larger oval shows
the area of one half inch barometric depression. The arrows
show the observed directions of the wind over an area about
900 miles in diameter, and this storm was nearly stationary for
four days.

272. *American Storms.*—Figure 60 represents the winds as observed near the centre of a violent storm of rain and snow, which

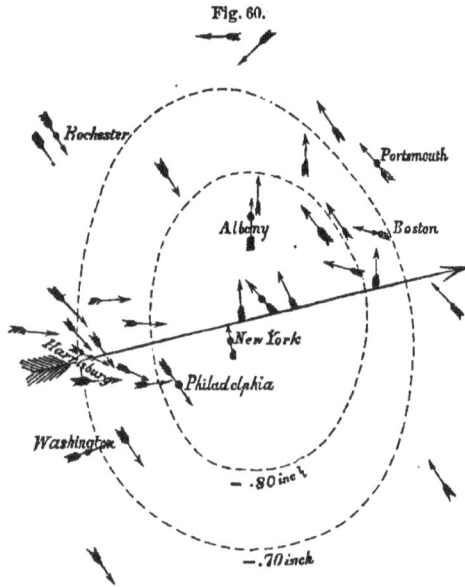

Fig. 60.

was experienced in
the neighborhood of
New York, on the
16th of February,
1842. The small
oval line shows the
area within which
the barometer sunk
eight tenths of an
inch below the mean,
and the larger oval
shows the area of
seven tenths inch
barometric depres-
sion. The long ar-
row represents the
direction in which
the storm advanced.
The short arrows
show the observed directions of the wind over an area about 500
miles in diameter.

The principles already stated are more fully illustrated by Plate
III., which represents the principal phenomena of a storm which
passed over the United States in December, 1836. The upper
map represents the phenomena for 8 P.M., December 20, and the
lower map represents the same for 8 A.M., December 21. A

comparison of the two maps shows the progress of the storm in twelve hours. The area of the rain or snow is represented by the dark shade near the middle of each map, and the lighter shade on the margin of the rain represents the region where clouds prevailed without rain. Throughout the remaining portion of the United States, as far as the maps extend, clear sky prevailed.

The dotted curve lines represent the state of the barometer. On map first the inner curve shows the area where the barometer was depressed four tenths of an inch below the mean; the next curve shows where the barometer was two tenths of an inch below the mean; the next curve shows the barometer at its mean height, while farther eastward the barometer stood two tenths of an inch and four tenths of an inch above the mean. On map second these curves are seen to have been somewhat modified in form, and to have been carried eastward a distance of about 450 miles.

The arrows show the directions of the wind as actually observed at a large number of stations, and these directions will be seen to conform generally to the principle stated in Art. 270, with a few exceptions, which may, perhaps, be ascribed to the influence of local causes.

273. *Distinction between the Direction of the Wind and that of the Storm's Progress.*—It will thus be seen that the direction of the wind at any place is entirely distinct from that of the storm's progress over the earth's surface. While the storm advances steadily eastward, the wind has every possible direction at different places within the limits of the storm.

At places on the north side of the centre of a great storm the wind generally sets in from the north of east as the storm approaches, and as the storm passes by the wind changes to the northwest, veering round by the north point. At places on the south side of the centre of the storm the wind generally sets in from the south of east as the storm approaches, and as the storm passes by the wind changes to the southwest, veering round by the south point.

Frequently the centre of a great winter storm is situated beyond the limits of the United States on the north, and then, throughout the entire United States, as far as observations have extended, the wind blows from the E. or S.E. on the front of the storm, and from the W. or S.W. on the rear of the storm.

274. *Lull at the Centre of a Storm.*—Near the centre of a great storm there is generally a lull of the wind, and sometimes a calm. Sometimes the clouds open, exhibiting considerable clear sky, and occasionally the clouds disappear entirely for several hours, exhibiting a clear sky, with little wind and a mild temperature. Soon after the centre of the storm has passed eastward of the observer the wind generally changes to the west, and the barometer begins to rise. The rain or snow, which may have been temporarily suspended, is renewed, generally with considerable violence, which, however, in such cases, is not usually of long continuance.

275. *Wind on the Extreme Borders of a Storm.*—Near the line of maximum pressure which surrounds a violent storm there is generally but little wind, and on each side of that line the winds are irregular in their direction, but generally tend outward from the line of greatest pressure. Hence it happens that near the extreme borders of a storm the winds are found blowing in nearly opposite directions, on one side inward toward the storm, and on the other side outward from the storm.

276. *How Winds are Propagated from Place to Place.*—Since on the opposite sides of a storm the wind blows in nearly opposite directions, while the entire storm makes progress toward the east, it is evident that some winds must be propagated from place to place nearly in the *same* direction as that in which they blow, while others are propagated in a direction *opposite* to that in which they blow. When a great storm springs up near the Mississippi, the wind at St. Louis is generally easterly, while throughout New York and Ohio the wind is from the west. Subsequently this easterly wind is felt at Cincinnati, then at Pittsburg, and afterward at New York; while the entire storm is traveling steadily eastward; that is, the easterly wind is propagated from St. Louis to New York in a direction *opposite* to that in which the wind blows.

After the centre of the storm has passed, a west wind springs up at St. Louis, and this west wind is felt successively at Cincinnati, Pittsburg, and finally at New York, having been propagated in the *same* direction as that in which the wind blows. The former wind is said to be propagated by *aspiration*, the latter by *impulsion*, as stated in Art. 142.

277. *Temperature near the Centre of a Storm.*—During an extensive rain-storm the temperature of the air generally rises above its mean height for that season of the year. This increase of temperature frequently amounts to 10° or 20°, and sometimes even 30°. This is caused by the latent heat which is liberated from the vapor when it is condensed into water. The centre of the area of high thermometer frequently does not coincide with that of the area of low barometer, or with the centre of the area of rain and snow. In the United States, on the northeast side of a storm, at a distance of over 500 miles from the area of rain and snow, the thermometer sometimes rises even 20° above its mean height. It seems probable that the heat which is liberated in the condensation of the vapor expands the upper portion of the atmosphere, and is drifted eastward far in advance of the storm.

278. *Low Temperature succeeding a Storm.*—As the heated air rises, the cold upper air descends to take its place, and the storm is suddenly succeeded by a temperature 10 or 20 degrees *below* the mean. Thus, when a storm is prevailing in the middle of the United States, the lowest temperature of the month may occur at St. Louis on the same day that the highest temperature occurs at New-York.

279. *Course of Storms modified by Local Causes.*—Local causes which tend to produce an upward current of the air exert an influence upon the course of storms. High mountain peaks are of this description. The storms of Europe are very much modified, and sometimes in a great measure controlled, by the Alps of Switzerland. By the interposition of these mountains, the air which sweeps over them is forced up to a great height, where it is suddenly cooled; its vapor is condensed; heat is thereby liberated, by which the surrounding air is expanded, and rises above the usual limit of the atmosphere. It thence flows off laterally, leaving a diminished pressure beneath the cloud; that is, the barometer shows a diminished pressure in the neighborhood of the mountain.

K

The mountain thus becomes the centre of a great storm, and the storm may continue stationary for several days, being apparently held in its place by the action of the mountain.

280. *Influence of the Gulf Stream.*—The Gulf Stream also gives rise to upward currents of the atmosphere. The Gulf Stream is a hot river, which comes out of the Gulf of Mexico, and sweeps round the southern part of Florida, whence it proceeds in a course nearly parallel to the coast of the United States, and distant from it about 100 miles. Its temperature in lat. 40° is generally 20° warmer than the surrounding ocean. The air over the Gulf Stream is warm and highly charged with vapor. Over this stream ascending currents are continually forming, and storms are more frequent in its neighborhood than in other parts of the ocean.

Moreover, if a storm commences any where in the vicinity of the Gulf Stream, it naturally tends toward this stream, because here is the greatest amount of vapor to be precipitated; and when a storm has once encountered the Gulf Stream, it continues to follow that stream in its progress eastward, so that most of the storms which prevail on the coast of the United States have their centre over the Gulf Stream, and follow the path of this stream in its progress eastward.

281. *Theories of Redfield and Espy.*—In recent times the study of Meteorology has been greatly promoted by the labors of Messrs. Redfield and Espy. Mr. Redfield maintained that in great storms the air moves in circles round the centre, while Mr. Espy maintained that the *tendency* of the wind is in the direction of radii toward the centre, and that the actual motion of the wind is inward toward the centre.

Observations have shown that an exactly circular motion of the wind rarely, if ever, occurs, and also that the air never moves exactly in the direction of radii toward the centre of the storm; but in almost all violent storms we find a combination of these two movements, viz., a pressure of the air inward, and a tendency to circulate round the centre, so that the actual motion of the wind seems to be spirally inward toward the centre; and here we must suppose the air to escape by rising upward from the earth's surface, and spreading out in the upper regions of the atmosphere.

282. *Cause of the Low Barometer near the Equator.*—The same principles which are developed in the action of storms are exemplified on a grand scale in the general circulation of the atmosphere. · The N.E. and S.E. trade winds, encountering each other near the equator, are forced up to a great height, where their vapor is condensed; copious rain follows; by the liberated heat the air is expanded, and flows off laterally from above. This causes the barometer to fall at the equator, and to rise at some distance on each side of the equator.

283. *Low Barometer near Lat. 64°.*—In like manner, near the parallel of 64° northerly and southerly winds encounter each other, producing, also, an abundant precipitation, with a low barometer near this parallel, and a high barometer at some distance on each side of this parallel.

Abundant rains, then, near the equator and the parallel of 64°, are the cause of the low barometer near those parallels, and they are also, in part, the cause of the high barometer near the poles and the parallel of 32°.

284. *Cause of the Uniformity of the Monsoons.*—The uniformity and strength of the S.W. monsoon in India, described in Art.152, is due to the vast amount of vapor precipitated on the Himalaya Mountains. The heat which is liberated in this condensation causes the air over the mountains to expand and flow off in the higher regions of the atmosphere, causing a greatly diminished pressure in the lower atmosphere, and this cause converts the S.W. wind of India, which otherwise might be a feeble and variable wind, into a strong and permanent wind throughout the warmer months of the year.

SECTION II.

CYCLONES.

285. *Cyclones defined.*—The inequalities of the earth's surface, especially in hilly countries, greatly modify the direction of the wind, so that in great storms the movements of the atmosphere often seem very complex and anomalous. Over the ocean these disturbing causes do not exist, and here we find that in violent storms the movements of the air are much more regular and uniform. This motion of the wind has generally been found to be in great circuits, spirally inward toward the centre of the storm, and such storms are now commonly designated by the term *cyclone*. These storms prevail in the neighborhood of the West India Islands, where they have long been known by the name of hurricanes. They are also common in the China Sea and in the Indian Ocean, on both sides of the equator.

286. *Season of Cyclones.*—In the West Indies, cyclones are almost exclusively confined to the months from July to October, being most common in the month of August; also, in the China Sea and the Bay of Bengal they are most prevalent at about the same period of the year. In southern latitudes they are most common from January to March.

287. *Where do Cyclones originate?*—There is no instance on record of a hurricane having been encountered on the equator, nor of any one having crossed that line, although two have been known to rage at the same time on the same meridian, but on opposite sides of the equator, and 10° or 12° apart. They originate near the equatorial limit of the trade winds, where these winds are irregular. The West India hurricanes generally originate between lat. 10° and 20° N., and long. 50° and 60° W., on the borders of the zone of calms and variable winds, which corresponds with the zone of constant precipitation of rain.

288. *Paths of Cyclones.*—In the northern hemisphere, during the early part of their course within the region of the trade winds, cyclones travel toward the west, inclining somewhat toward the north. Near lat. 20° the motion from the equator is more decided, and in lat. 25° their motion is about N.W. Near the parallel of 30° their course is almost exactly north, and soon they begin to veer toward the east, after which their motion is nearly parallel to the coast of the United States. Several storms have been traced from lat. 10° or 15° up to lat. 45° or 50°, and the path of the centre of greatest violence bears some resemblance to a parabola, of which the most westerly point lies near the parallel of 30°. This path is represented by the line ABC, Fig. 61.

In the southern hemisphere cyclones pursue a similar course. Commencing near the equator, they advance at first only a little

Fig. 61.

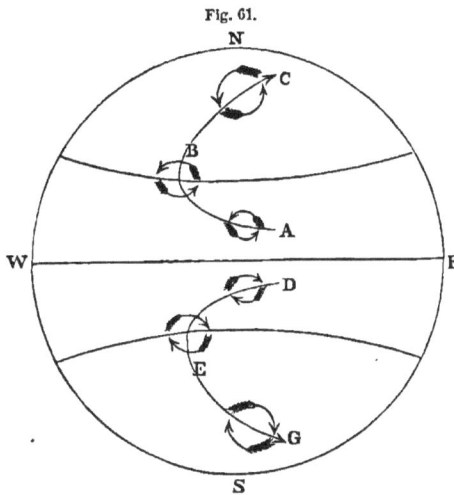

south of west. This southerly motion increases until near lat. 26°, when the motion is exactly toward the south, after which they gradually veer toward the southeast, the entire path, DEG, forming a curve which is almost perfectly symmetrical with that of cyclones in the northern hemisphere. The latitude where the path of the cyclone changes from west to east coincides nearly with the polar limit of the trade winds.

289. *Gyratory Movement of Cyclones.*—The air in cyclones has not merely a movement of translation, but also a gyratory motion about the centre of the storm. The motion of the air is spirally inward, as has been already shown in the storms of the United States, but over the ocean the whirling motion is usually more decided than it is over the land. North of the equator this gyratory motion is from right to left, or in a direction *contrary* to that of the hands of a watch. South of the equator the motion is from left to right, or in the *same* direction as that of the hands of a watch.

Near the centre of the hurricane there is generally a great fall of rain, which is usually accompanied by the most magnificent displays of thunder and lightning.

290. *Rate of Motion.*—The rate at which cyclones travel is very variable. In the West India cyclones the highest rate which has been observed is 43 miles per hour, and the least 10 miles per hour; the mean being 26 miles. In the Bay of Bengal the observed rate varies from 2 to 39 miles per hour, and in the China Sea from 7 to 24 miles per hour. In the South Indian Ocean the observed rate varies from 1 to 10 miles per hour. Some cyclones travel so very slowly that they may almost be considered stationary.

The direction and velocity of the wind are, however, entirely distinct from those of the storm's progress. While the storm sometimes advances at the rate of less than 10 miles per hour, the velocity of the wind may exceed 100 miles per hour.

291. *Diameter of Cyclones.*—Cyclones extend over a circle from 100 to 500 miles in diameter, and sometimes 1000 miles. In the West Indies they are sometimes as small as 100 miles in diameter, but on reaching the Atlantic they dilate to 600 or 1000 miles. Sometimes; on the contrary, they contract in their progress, and while contracting they augment fearfully in violence. The violence of the wind increases from the margin to the centre, with the exception of a limited space exactly at the centre, where the atmosphere is frequently quite calm.

292. *Premonitions of a Cyclone.*—Previous to the commencement of a cyclone the air is observed to be close, sultry, and oppressive, and the wind is moderate or calm. A fresh breeze sets in from the east, and rises and falls with a moaning sound; after a few hours it is succeeded by a lull, which may last for an hour or more, after which the wind changes to the west, often with great suddenness, and blows with increased violence, and this is usually the time of greatest danger to vessels.

The approach of a cyclone is often announced by a swell of the ocean, resulting from the action of the wind upon a neighboring sea, while the waves thus excited advance more rapidly than the storm.

During the passage of the cyclone the barometer oscillates in a remarkable manner, rising and falling rapidly, so that a great barometric oscillation almost always announces the approach of a tempest. The most rapid fall begins from three to six hours before the passage of the centre. The barometer is lowest near the middle of the storm area, and begins to rise before the strength of the cyclone is over.

The fall of the barometer during the passage of the cyclone varies according to the intensity of the storm. It frequently amounts to one inch, and has been known to exceed two inches. The rise of the barometer after the storm is usually as rapid as was its fall on the approach of the storm. See Table XXXIII.

293. *Duration at any Place.*—The duration of the storm at any place depends upon the extent of the storm, and the velocity with which it advances. If the storm be only 100 miles in diameter, and advances 20 miles per hour, its duration at any place can not exceed five hours. If the diameter of the storm be greater, or its progress less rapid, its duration at a given place will be increased.

294. *Cause of the Parabolic Course of Storms.*—The parabolic course of storms from near the equator toward the poles results from the rotary motion of the earth. When a large mass of air in the northern hemisphere is put in rotation about a vertical axis, the particles on the east side of the centre, crossing successively parallels of latitude whose easterly motion is less than their own, are deflected toward the east; that is, toward the right. So, also, the particles on the west side of the centre, crossing successively parallels of latitude whose easterly motion is greater than their own, are deflected toward the west, which is also toward the right. Particles on the north or south side of the centre are deflected in a similar manner; that is, the particles of the revolving mass of air, in every portion of their circuit, are deflected toward the right. Hence on the equatorial side of the revolving mass of air there is a tendency toward the equator, while on the polar side there is a tendency toward the pole. Now this deflecting force increases from the equator toward the pole, being proportional to the sine of the latitude. Hence the pressure on the polar side toward the pole is greater than on the opposite side toward the equator, and the revolving mass accordingly moves in the direction of greatest pressure; that is, toward the pole.

Within the limit of the trade winds the revolving mass is carried westward by the general westward motion of the atmosphere, while it is crowded northward by the force just described, so that the actual progress of the storm is toward the north of west. After escaping from the trade winds, the general motion of the atmosphere carries the storm eastward, while the force just described urges it northward; that is, the actual progress of the storm is toward the north of east.

By a similar course of reasoning, the parabolic path of cyclones in the southern hemisphere may be explained.

SECTION III.

TORNADOES.

295. Sometimes near the centre of a great storm the general inward tendency of the air causes a violent whirlwind, or tornado, where the wind revolves with such violence as to prostrate the largest trees, demolish buildings, and transport heavy bodies to a great distance. Such a whirlwind occurred in Northern Ohio February 4, 1842, near the centre of an uncommonly severe storm of rain. In this tornado large buildings were lifted entire from their foundations, carried a distance of several rods, and then dashed to pieces. The fragments were strewed all along the track, and some were carried a distance of seven or eight miles. Large oak-trees, two feet in diameter, were snapped off like reeds, and others were so twisted as to be reduced to a mass of splinters not much thicker than a man's finger. The breadth of the track did not much exceed half a mile, and the most destructive portion was still more limited. The duration of the tornado at one place did not much exceed one minute. The tornado advanced over the earth, in a direction N. 33° E., with a velocity of 34 miles per hour.

296. *Tropical Tornadoes.*—Similar tornadoes occur within the tropics, and here exhibit even greater violence than they do in the United States. In the great tornado which passed over Barbadoes in 1780, the strongest buildings were entirely demolished; the largest trees were torn up by the roots; a 12-pounder gun was moved a distance of 140 yards; a multitude of ships were wrecked, and over 4000 persons perished.

In a hurricane which occurred in June, 1822, near the mouth of the Ganges, a vast amount of property was destroyed, and upward of 50,000 persons perished, chiefly from the inundation of the rivers.

297. *Effects of Tornadoes.*—The motion of the air in tornadoes is spirally inward and upward, so that from each side of the track objects are drawn inward toward the centre of the track, and very heavy bodies are carried up in the centre. Light objects are elevated high into the air, and are sometimes carried

Fig. 62.

many miles before they are thrown out of the vortex.

Fig. 62 represents a portion of the track of a tornado which passed over New Haven in 1839. The tornado advanced in a direction N. 50° E. On the right-hand side of the track the prostrate trees were uniformly inclined toward the north, while on the left-hand side many of them were inclined toward the south.

Tornadoes are uniformly preceded by an unusual heat; they are invariably accompanied by lightning and rain, and frequently by hail.

When a tornado passes over a hilly country, it sometimes rages with destructive violence on the hill-tops, while objects in the intermediate valleys are entirely uninjured, showing that a violent whirlwind may prevail at a moderate elevation, but without reaching the earth's surface.

298. *Appearance of Explosion.*—When a violent tornado passes over a building where the doors and windows are closed, the walls are sometimes thrown outward with great force, the house presenting•the appearance of an explosion, indicating that the pressure of the air on the outside of the building was suddenly diminished, and the house was burst open by the expansion of the air within.

SECTION IV.

PILLARS OF SAND, AND WATER-SPOUTS.

299. Tornadoes are probably similar to the small whirls which are often seen in the streets, especially on dry and calm days of spring or summer, and which raise up a dense column of dust, even to the tops of the houses. In these whirls the motion of the air is spirally inward and upward, so that light objects in their vicinity are sucked into the vortex, and carried up to the top of the whirl, where they escape laterally, and descend at some distance on either side. These small whirls sometimes revolve from left to right, and sometimes from right to left, while in the northern hemisphere large whirlwinds, several miles in diameter, always revolve from right to left. ' The whirls seen in our streets are sometimes only a few inches in diameter, but sometimes in the open fields they occur several feet in diameter, and carry up leaves of trees and light objects of considerable size.

On the deserts of Africa similar whirls often raise vast pillars of sand, which sometimes prove fatal to entire caravans. Bruce states that in Abyssinia he beheld eleven vast columns of sand moving over the plain at the same time. Similar whirls are of common occurrence in India.

300. *Whirlwinds caused by Fires.*—These whirls may be set in motion by whatever causes a strong upward motion of the air. An extensive fire frequently produces this effect. When large fires are burning on the Western prairies, violent whirls are frequently formed, having a force sufficient to lift a man from the ground and transport him to a considerable distance. At such times the flame is sometimes collected into a fiery column, rising to the height of 200 feet or more.

Some years since, during the burning of a canebrake in Alabama, several whirls were formed in the midst of the flames, some of which rose to the height of 200 feet, and in form resembled the upper cone of an hour-glass.

Similar effects were produced by the conflagration of Moscow, September 14–20, 1812.

301. *Water-spouts.*—When a violent whirl is formed over water, considerable spray is raised from the surface of the water, and this spray is carried up in the centre of the whirl, presenting the appearance of a dense solid column. This phenomenon is called a water-spout. Water-spouts are of variable dimensions, but sometimes they attain a diameter of several rods, and a height of half a mile.

These whirls generally form, in the first instance, at a considerable height in the air, and do not reach down to the surface of the sea. If there is a low cloud over it, the under surface of the cloud is rolled into a conical form. This inverted cone seems attached to the cloud, and sometimes becomes rapidly elongated. Sometimes it swings backward and forward, coils up, and disappears, and the spout is not completed; but at other times it gradually extends so as to reach down to the surface of the water. As the column approaches the surface of the sea, the latter becomes violently agitated, and the spray is whirled round with a rapid motion. The spout now forms a continuous column, extending from the water to the cloud, and often resembles a large elephant's trunk dangling from the clouds. Its color is generally of a sombre gray, like that of the clouds, but sometimes it appears · black, like a dense smoke.

This spout has both a rotary and a progressive motion. The whirling motion extends to but a moderate distance around the column, and beyond this there prevails a calm. The phenomenon lasts but a short time. After a few minutes the trunk contracts so as no longer to reach the surface of the sea, the black cloud draws itself up, and the trunk gradually disappears. Sometimes the spout commences with the rising of spray from the surface of the water, which gradually ascends until the column is complete from the water to the clouds. When the spout is complete, there is heard a roaring noise like that of a great waterfall. ·

Fig. 63.

Subsequently the cloud sometimes discharges itself in a heavy rain, and this rain is never salt, even in the open ocean, showing that this water was precipitated from the clouds, as in ordinary rains. Fig. 63, on the preceding page, shows a water-spout in three stages of its progress. First, the column is incomplete; next, the column is entire; and, finally, the smoky aspect of the column disappears, and the column begins to break up.

Water-spouts generally form during a period of great heat, and are most frequent in the calm regions between the tropics. Two or three of these spouts are sometimes formed simultaneously, proceeding from the same cloud. In May, 1820, on the edge of the Gulf Stream, seven water-spouts were seen in the course of

Fig. 64.

half an hour. Fig. 64 represents a water-spout seen in 1858 on the River Rhine.

302. *Showers of Toads, Fishes, etc.*—During violent storms showers of small animals sometimes descend from the sky. M. Peltier, of France, states that he once saw a multitude of small toads descend to the earth. They fell upon his hat, upon his hands, and the ground about him was covered with them. Several observers in France, in India, and elsewhere, have seen showers of small fish descend from the sky.   Others have observed showers of sand, of straws, etc.

These phenomena are explained by supposing that the objects mentioned were elevated from the earth in a violent whirl, which transported them to a considerable distance, and then dropped them upon the earth.

In 1833, near Naples, a whirlwind passed over an orange-grove, and a multitude of oranges were carried up in the whirl.   Some minutes afterward a shower of oranges fell upon a roof at a considerable distance.

In 1835, in France, the water of a small pond containing a large quantity of fish was drawn off by a whirlwind. . These animals may have been transported a distance of many rods, perhaps several miles, but they must ultimately. have fallen to the earth, furnishing a shower of fishes.

SECTION V.
PREDICTIONS OF THE WEATHER.

303. The character of the weather at any place is affected by so many circumstances which may transpire at distant parts of the world, and which can be but very imperfectly known to us, that it is impossible to predict, except very imperfectly, what weather may be expected at a given time and place.   To a limited extent, however, such predictions are possible.

304. *Predictions founded upon the Constancy of Climate.*—Relying upon the constancy of climate, which has been established by observation, we may predict the probable general character of any month of the year.

The climate of a country remains permanently the same from age to age.   Observations continued for an entire century at various places in the United States and Europe indicate no change in the mean temperature of the year, or that of the separate months; no change in the range of the thermometer; no change in the time of the last frost of spring or the first frost of autumn; in the annual amount of rain or snow, or in the mean direction of the wind.   It is not certain that the climate of any country, in either of these respects, has changed appreciably in 2000 years. By the destruction of forests, the earth is more directly exposed to the rays of the sun; the moisture of the ground is more readily evaporated; streams more frequently dry up in summer, and droughts become more frequent and severe. But these changes do not seem to affect in a sensible manner the mean temperature of any place, or the annual amount of rain.

Assuming, then, the established constancy of climate, we can predict beforehand the probable character of any month of the year. Thus, at New Haven, the probable mean temperature of any future January will be 26°. We may be tolerably sure that it will not be higher than 36°, nor lower than 17°. The thermometer in January will never rise above 64°, nor sink below —24°. The entire annual amount of rain at New Haven will not exceed 55 inches, and will not be less than 34 inches.

305. *Conclusions drawn from anomalous Months.*—Moreover, if several months in succession have been unusually warm or unusually cold, instead of concluding that the climate has permanently changed, and that the succeeding months will be similar in character, we should rather anticipate months of the opposite description, since the mean temperature of the year fluctuates within very narrow limits, and the longer a period of unusually warm weather continues, the greater is the probability that the succeeding months will be unusually cold. Predictions of this kind are legitimate deductions from scientific data.

306. *Predictions founded upon the established Laws of Storms.*— Since great storms have been found to observe pretty well defined laws, both as respects the motion of the wind and the direction of their progress, we may often recognize such a storm in its progress, and anticipate changes which may succeed during the next few hours. When it is possible to obtain telegraphic reports of the weather from several places in the Valley of the Mississippi and its tributaries, we may often predict with confidence the approach of a great storm twenty-four hours before its violence is felt at New York.

307. *Observations of the Meteorological Instruments at a single Place.*—When we are restricted to observations at one locality, our predictions of the weather must needs be more uncertain, and the conclusions to be derived from a motion of meteorological instruments are not the same for all parts of the world. Along the Atlantic coast of the United States the approach of a violent N.E. storm is generally indicated by the barometer rising above its mean height; at the same time the wind veers to the N.E., and the atmosphere grows hazy. After the rain or snow commences, the barometer begins to fall; when the barometer reaches its lowest point, the wind changes to N. or N.W., after which the barometer begins to rise.

If a gale sets in from the E. or S.E., and the wind veers by the S., the barometer will continue falling until the wind becomes S.W., when a comparative lull may occur, after which the gale will be renewed, and the change of the wind toward the N.W. will be accompanied by a fall of the thermometer, as well as a rise of the barometer.

A considerable and rapid depression of the barometer—for instance, a fall of three fourths of an inch in twenty-four hours—indicates an approaching storm, with rain or snow. The wind will be from the northward if the thermometer is low for the season, from the southward if the thermometer is high. If the barometer falls with a rising thermometer and increased dampness, wind and rain may be expected from the southward.

A rapid rise of the barometer indicates unsettled weather; a slow rise indicates fair weather. The result of all rapid changes in the weather, or in any of the instrumental indications, is brief in duration, while that of a gradual change is more durable.

308. *Prognostics from the Clouds, Face of the Sky, etc.*—When the upper clouds move in a direction different from that of the lower clouds, or that of the wind then blowing, they foretell a change of wind.

When the outlines of cumulus clouds are sharp, it indicates a dry atmosphere, and therefore presages fine weather. Small inky-looking clouds foretell rain. A light scud driving across hazy clouds indicates wind and rain.

Remarkable clearness of the atmosphere near the horizon, and an unusual twinkling of the stars, indicate unusual humidity in the upper regions of the atmosphere, and are therefore indications of approaching rain.

Halos, coronæ, etc., presage approaching rain or snow.

Dew and fog are indications of fine weather.

CHAPTER VII.

ELECTRICAL PHENOMENA.

SECTION I.

ATMOSPHERIC ELECTRICITY.

309. *Means of observing the Electricity of the Atmosphere.*—The atmosphere is almost always charged with electricity, and this electricity exerts an important influence upon various meteorological phenomena.

In order to observe this electricity, an insulated conductor should be elevated to a considerable height above the earth. At the Observatory of Kew, near London, a tube of thin copper, 16 feet high, and surmounted by platinum points, is supported by a cylinder of glass placed under the dome at the top of the Observatory. The copper tube passes through the top of the dome without touching it, and the rain is excluded from this opening by an inverted copper dish fitted to the tube. This copper tube may be made to communicate at pleasure with the electrometers.

310. *Electrometers.*—The most common electrometer is Volta's.

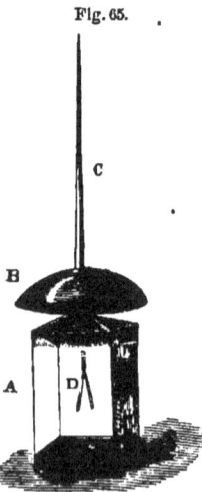

Fig. 65.

This consists of two straws, D, Fig. 65, two inches in length, suspended by hooks of fine copper wire, and at a distance of one twentieth of an inch from each other, and covered by a glass jar, A. When the two straws are similarly electrified they recede from each other, and the intensity of the charge is indicated by the amount of the divergence. This divergence is measured by an ivory scale graduated to twentieths of an inch. B is a metallic dish to protect the electrometer from the rain, and C is a pointed conductor for collecting the electricity.

It is desirable to have a series of electrometers for measuring electricity of different degrees of intensity. For the feeblest electricity the gold-leaf electrometer may sometimes be employed; and when the electricity is very intense it is important to have an instrument for measuring the length of the spark. This may consist of a sliding rod terminated by a brass ball, which can be set at any distance from the insulated conductor.

311. *Electricity at considerable Elevations.*—The electrical condition of the higher strata of the air has been ascertained by means of kites and balloons. When a kite is used for this purpose, the string should be wound with fine wire in order to make it a conductor of electricity, and the kite must be insulated by attaching the lower end of the string to some non-conductor such as silk or glass.

Small balloons are sometimes employed for the same purpose, and a conducting cord connects the balloon with an electrometer near the earth's surface.

By instruments like these it is found that the air is generally charged with positive electricity, but it is subject to great variations of intensity, and clouds are frequently charged with negative electricity.

312. *Diurnal Variation of Electricity.*—The intensity of atmospheric electricity is found to vary with the hour of the day. From the mean of three years' observations made at Kew, it appears that at 4 A.M. the electric tension is represented by 20 on Volta's electrometer; from this hour the electricity increases to 10 A.M., when it is represented by 88; from that time it decreases to 4 P.M., when it is represented by 69; it then increases to 10 P.M., when it is represented by 104; from which time it decreases till 4 A.M.; that is, there are two daily maxima of intensity and two daily minima.

313. *Monthly Variation of Electricity.*—The intensity of atmospheric electricity also varies with the season of the year. At Kew, the mean electric tension is least in June, remaining nearly the same through the summer months, after which the electricity increases steadily till January, continuing nearly the same through February, after which it decreases till the next June; that is, there is one annual maximum of intensity and one minimum.

At Brussels, also, the maximum occurs in January and the

L

minimum in June, while at Munich the maximum occurs in December and the minimum in May.

At Brussels the electric tension in winter is nine times as great as in summer; at Kew it is six times as great; and at Munich it is only twice as great in winter as in summer.

314. *Variations with the Altitude.*—The intensity of atmospheric electricity increases with the altitude above the surface of the earth. This law has not been fully verified for elevations exceeding 100 feet. Experiments with electric kites have obtained signs of electricity the more powerful as the kite rose to a greater elevation. Experiments of this kind have been carried to the height of 810 feet.

Similar results have been obtained by means of an arrow projected into the air, the arrow being provided with a conducting wire whose extremity communicated with a straw electrometer.

Gay-Lussac, during his aerial voyage in 1804, suspended from his balloon a wire 170 feet long, and connected the upper end with an electrometer. This experiment indicated that the electricity of the air was positive, and increased with the altitude.

In a balloon ascent in 1862, Mr. Glaisher found that the air was charged with positive electricity, but becoming less and less in amount with increasing elevation, till at the height of 23,000 feet the amount was too small to measure.

315. *Electricity in cloudy Weather.*—When the sky is covered with clouds, the electricity is subject to frequent changes of kind as well as intensity, being sometimes positive and sometimes negative. The electricity is seldom negative except when rain is falling. During a snow-storm the lower strata of the air exhibit electricity of great intensity.

During the passage of a thunder-shower, the electricity frequently changes in two or three minutes from positive to negative, and then back again to positive, sometimes half a dozen of these changes occurring in a single shower. The electricity also at such times has great intensity, and sparks are sometimes obtained from the conductor more than an inch in length, giving a severe shock when passed through the human system.

316. *Is Atmospheric Electricity the result of Friction?*—Philoso-phers are by no means agreed as to the origin of atmospheric electricity. Friction is one of the most common sources of elec-tricity. Dry air rubbing against dry air, or any other substance, develops little if any electricity; but moist air rubbing against the surface of the earth acquires positive electricity.

In violent tornadoes we uniformly observe electricity of great intensity. This may be due in part to the friction of the air upon the earth. But we can not consider friction to be the principal source of atmospheric electricity, because there is no uniform relation between the force of the wind and the intensity of the electricity.

317. *Is it the result of Combustion?*— Combustion is another source of electricity. When coal is burning, the carbonic acid gas which escapes is positively electrified, while the coal has negative electricity. The atmosphere, therefore, must receive some electricity from the combustion which takes place on the surface of the earth; but this cause must be entirely inadequate to account for the enormous quantities of electricity exhibited in thunder-showers.

318. *Is it the result of Vegetation?*—Vegetation is a source of electricity. During the day, plants give out oxygen which is charged with negative electricity; and during the night they give out carbonic acid gas, which is charged with positive elec-tricity. These two processes in a measure neutralize each other.

319. *Is it the result of unequal Temperature?*—The unequal tem-perature of the different parts of the earth has been supposed to be a source of atmospheric electricity. There are several metals which develop electricity when brought in contact and unequal-ly heated. In some of the mines of England, currents of elec-tricity have been detected within the earth, and these currents have been ascribed to a varying temperature acting upon the heterogeneous materials of the earth.

This cause may explain permanent currents existing in the earth, but does not seem adequate to account for the enormous quantity of free electricity which often exhibits itself in thunder-showers.

320. *Is it the result of sudden Condensation of Vapor ?* — Since atmospheric electricity is feeble before the formation of a storm, and rapidly attains its maximum during a thunder-storm, it has been supposed that electricity is liberated in the act of condensation of the vapor of the air.

When the steam issuing from the boiler of a steam - engine is suddenly condensed, a great amount of electricity is liberated. But it is claimed that this electricity is not due to simple condensation, but to the friction of the condensed particles against the sides of the orifice through which the steam escapes.

321. *Is Atmospheric Electricity due to Evaporation ?*—Evaporation is probably the principal source of atmospheric electricity. The following experiment shows the production of electricity by evaporation. If upon the top of a gold-leaf electrometer we place a metallic vessel containing salt water, and drop into the water a heated pebble, the leaves of the electrometer will diverge. The vapor which rises from the water is charged with positive electricity, while the water retains negative electricity.

The water used in this experiment must not be perfectly pure, but must contain a little salt, or some foreign matter. The evaporation of the water of the ocean must therefore furnish a large amount of electricity; and fresh water must also furnish some electricity, for the water of the earth is never entirely pure.

322. *Diurnal change of Electricity explained.*—The diurnal variation in the intensity of atmospheric electricity is to be ascribed partly to real changes in the amount of electricity present in the air, and partly to variations in the conducting power of the air.

Just before sunrise the electricity has a feeble intensity, because the moisture of the preceding night has transmitted to the earth a portion of the electricity which was previously present in the air. After the sun rises, new vapor ascends and carries with it positive electricity, so that the amount of electricity in the air increases. Toward noon the air becomes dry, and transmits less readily the electricity accumulated in the upper regions of the atmosphere; so that, although the amount of electricity in the air is continually increasing, an electrometer near the earth's surface indicates an apparent diminution. Toward evening the air grows cool, again becomes humid, and transmits more readily to the earth the electricity accumulated in the upper regions of the atmosphere. The effect produced upon an electrometer therefore increases until some hours after sunset; but since during the night there is a constant discharge of electricity from the air to the earth, the electrometer soon indicates a diminished intensity, which continues until toward morning.

323. *Monthly change of Electricity explained.*—The same principle explains why the electricity of the air appears less intense in summer than in winter. In summer the air is warm and dry, and opposes more resistance to the flow of electricity from the higher regions of the atmosphere, while in winter the moist air produces a contrary effect; so that, although the atmosphere doubtless contains more electricity in summer than in winter, it generally produces a less effect upon an electrometer placed near the earth's surface.

324. *Electricity developed in dry Houses.* — During the cold weather of a Northern winter, in houses which are kept quite warm and dry, and whose floors are covered with heavy woolen carpets, electricity is abundantly excited by simply walking to and fro upon the carpet. Sometimes in this manner there is developed electricity sufficient to give an unpleasant shock, and to ignite ether, gas, or other combustible substances. This electricity results from the friction of dry leather upon the woolen carpet, and it is prevented from escaping by the insulating power of the dry carpet, and the extremely dry floor of the building.

SECTION II
THUNDER-STORMS.

325. *How clouds become Electrified.* — We have found that the atmosphere ordinarily contains a large quantity of electricity. Since dry air is a non-conductor, the electrified particles in clear weather are in a measure insulated, and the electricity can not acquire much intensity; but when the vapor of the air is precipitated and a cloud is formed, the electricity, which was previously confined to the separate particles of the air, now finds a conducting medium more or less perfect, and it spreads itself over the surface of the cloud, thereby acquiring considerable intensity. It is generally admitted that the same quantity of electricity which exists in the cloud, existed in the air before the formation of the cloud, and that the cloud performs no other office than that of a conductor.

326. *Clouds negatively Electrified.* — A cloud thus electrified must necessarily have positive electricity, since in clear weather the electricity of the atmosphere is always positive. Such a cloud, when it approaches near another cloud having less electricity, or none at all, acts by induction upon the latter, decomposing its natural electricity, attracting the negative electricity and repelling the positive. The positive electricity thus repelled may sometimes be drawn off by near approach to another cloud, or to the earth, leaving only negative electricity upon the cloud. Hence probably result the frequent alternations of positive and negative electricity observed during a thunder-shower.

327. *Lightning.*—Two clouds having opposite electricities attract each other, and when the clouds come sufficiently near, the two electricities rush toward each other with great violence. This phenomenon is called *lightning*, and is accompanied by an explosive noise called *thunder.*

Since clouds are very imperfect conductors, when the electricity of one part of a cloud is discharged, the electricity of a distant part of the cloud is but slightly changed. Thus a single discharge does not establish a complete electrical equilibrium; but there is a change in the distribution of the electricities upon the surrounding clouds, and there must be a succession of discharges before the electricity is entirely neutralized. Hence results a succession of flashes of lightning and peals of thunder.

328. *Discharge of Electricity to the Earth.* — A cloud charged with electricity exerts an inductive influence upon the earth's surface immediately beneath it, decomposing its natural electricities, repelling electricity of the same kind, and attracting the opposite kind. Accordingly there will sometimes be a discharge of electricity from the cloud to the earth. This charge is usually received by the most elevated objects, such as mountains, hills, trees, spires, high buildings, etc. Trees are particularly exposed to strokes of lightning on account of their elevation, as well as of the moisture which they contain, and which renders them partial conductors of the electric fluid.

329. *Different forms of Lightning.*—Lightning exhibits a variety of forms, which have been designated by the terms zigzag, ball, sheet, and heat lightning.

Zigzag lightning presents a long, irregular, jagged line of light, like the ordinary spark drawn from an electric machine. This zigzag path is sometimes four or five miles, and perhaps even ten miles in length.

The irregularity of the path is ascribed to the compression of the air before the electricity, thereby opposing greater resistance, and turning the fluid aside to seek some path upon which the resistance is less.

330. *Ball Lightning.*—Ball lightning appears like a ball of fire, and is usually accompanied by a terrific explosion. It probably results from a charge of electricity unusually intense, which forces a direct instead of a circuitous passage through the air.

Some have supposed that ball lightning was the agglomeration of ponderable substances in a state of great tenuity, strongly charged with electricity.

331. *Sheet Lightning.*—Sheet lightning is a diffuse glare of light, sometimes illuminating only the edges of a cloud, and sometimes spreading over its entire surface.

This may sometimes be due to distant lightning which illumines a cloud, while the direct flash is hidden from the observer by intervening clouds. Sometimes it may result from a movement of electricity in the interior of a cloud which is a very imperfect conductor, producing an illumination analogous to that observed on a plate of moist glass employed in discharging an electrical machine.

332. *Heat Lightning.*—During the evenings of summer, the horizon is sometimes illumined for hours in succession by flashes of light unattended by thunder. This is called heat lightning. This illumination is sometimes due to the reflection from the atmosphere of the lightning of clouds so distant that the thunder can not be heard.

Sometimes, however, this light overspreads the entire heavens, showing that the electricity of the clouds escapes in flashes so feeble that they produce no audible sound. Such cases may occur when the air is very moist, the air being then a tolerable conductor, and offering just sufficient resistance to the passage of the electricity to develop a feeble light.

333. *Color of Lightning.*—The color of lightning varies from white to a rose color and violet. Zigzag lightning is generally white, sometimes of a purplish violet or bluish tinge. Diffuse flashes of lightning are often of an intense red, sometimes mixed with blue or violet.

These differences depend upon the density and moisture of the strata of air in which the clouds are formed, and also upon its conducting power. When the density of the medium is slight, the light becomes diffuse and reddish; when the density is considerable, the light is concentrated and brilliant.

Similar variations in the color of the electricity are perceived when the fluid is passed through a glass receiver in which the air has been rarefied by means of an air-pump.

334. *Duration of Lightning.*—The duration of ordinary flashes of lightning is less than the thousandth part of a second. This is proved by receiving the light of an electric discharge upon a white disc marked with black rays, when the disc is made to revolve with great rapidity. However great the velocity of rotation may be, the disc, when illumined by lightning, always appears stationary, showing that during the continuance of the illumination the disc had not revolved through any appreciable angle. If the disc were illumined for an instant by means of a lamp, by lifting and dropping a screen as suddenly as possible, the disc would appear of a uniform tint, and no separate rays would be seen.

335. *Cause of Thunder.*—Thunder is generally regarded as the result of the sudden re-entrance of the air into a void space, as in the experiment of a bladder tied over an open-mouthed receiver, and burst by the pressure of the external air. This vacuum is supposed to be generated by the lightning in its passage through the air. Electricity communicates a powerful repulsive force to the particles of air along the path of its discharge, producing thus a momentary void, into which immediately afterward the surrounding air rushes with a violence proportioned to the intensity of the electricity.

336. *Interval between the Flash and Report.*—Since the transmission of light is nearly instantaneous, while sound moves only 1100 feet per second, the sound will not reach the ear until some interval after the flash. By observing the interval between the flash and the report, the distance of the point where the discharge takes place can be computed. The longest interval mentioned by any observer is 72 seconds, indicating a distance of 15 miles. With the exception of this single instance, the longest interval recorded is 50 seconds, indicating a distance of 10 miles. This fact is very remarkable, since the noise of cannon may be heard to a much greater distance.

The average interval between the flash and the report is 12 seconds, and the shortest interval recorded is one second.

If we measure the angular height of the flash whose distance from the observer has been determined, we may compute the vertical elevation of the cloud above the earth.

337. *Duration of Thunder.*—Since a separate sound is produced

Fig. 66.

at each point along the entire line of the flash, and these points are generally at unequal distances from the observer, the sounds produced at different points of the line of discharge, though in fact simultaneous, reach the ear in slow succession. Thus an observer at A, Fig. 66, will first hear the sound resulting from the concussion at *a*, next at *c*, and finally at *b*. If *b* were 11,000 feet more remote than *a*, the first sound would be heard ten seconds before the last, and the thunder would be continuous for ten seconds.

The average duration of peals of thunder is 22 seconds, and the longest duration recorded is 56 seconds.

The prolonged duration of some peals of thunder is in part the effect of echoes.* In mountainous countries thunder peals are much longer continued, and the sound is more intense than in plane countries. This is due to the reflection of the thunder from the sides of the mountains in the same manner as the sound of a cannon is reflected. These echoes may also be produced by reflection of sound from clouds, as has been proved by the firing of cannon over the ocean.

338. *Rolling of Thunder.*—The variable intensity or rolling of thunder is due partly to the zigzag form of the discharge, in consequence of which there are frequently several different points of the flash which are equally distant from the observer; and the sounds produced at these points reach the ear simultaneously, producing the effect of a double or triple sound.

It is due in part to the unequal distance of different parts of the flash, the loudness of sound varying inversely as the square of the distance.

It may also be due in part to the fact that the electricity, in its long zigzag course, may pass through strata of air differing materially in density, which may result either from difference of elevation or difference in amount of moisture.

The rolling of thunder is also without doubt in a considerable degree the effect of echoes.

339. *Remarkable succession of Phenomena in Thunder.*—There is a certain succession of phenomena in thunder which occurs so frequently as to indicate that it is the result of a combination of circumstances of common, if not habitual occurrence. These phenomena occur in the following order:

1st. The flash of lightning.

2d. After an interval, generally of 10 or 12 seconds, the thunder begins with a rattling or rumbling noise, which increases, sometimes regularly, sometimes with vibrations, up to its maximum.

3d. Five or ten seconds after the first rumbling we hear a loud crashing sound, which sometimes continues for 5, 10, or even 20 seconds, and this again is succeeded by a rumbling noise, which gradually dies away.

Sometimes several maxima and minima succeed each other with great rapidity.

This circumstance of a crashing sound succeeding by a considerable interval the first rumbling of the thunder may perhaps be explained by the imperfect conducting power of the cloud. If we coat a Leyden jar with brass filings instead of tin foil, and charge it with electricity, upon discharging the jar in a dark room we find the light exhibits numerous ramifications, spreading out like branches from the trunk of a tree. A similar effect may be produced when electricity is discharged from a cloud. Let A B, Fig. 67, represent the zigzag discharge from one cloud to another, and

Fig. 67.

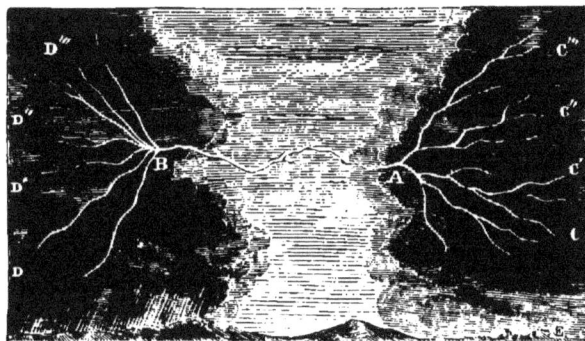

suppose the discharge of electricity from the interior of one cloud takes place by the branches A C, A C', etc., and from the interior of the other cloud by the branches B D, B D', etc. Then an observer at E would first hear the rattling sound resulting from the motion of the electricity along the paths A C, A C', etc., and this noise would not be of very great intensity. After a few seconds the sound of the concentrated discharge through A B will reach him, and he will hear a crashing noise, which will continue for several seconds with variable intensity. This will be succeeded by a low, rumbling noise, resulting from the partial discharge along B D, B D', etc., and this noise will be faint on account of the great distance.

340. *Height of Thunder Clouds.* — Thunder clouds are sometimes limited to a height of less than a quarter of a mile, and sometimes they rise to the height of at least three or four miles. Observers on the summit of hills less than a quarter of a mile in height, have seen thunder-showers below them, while they were

enjoying a cloudless sky.   On the other hand, La Condamine encountered a violent thunder-storm on a peak of the Cordilleras at
the height of 15,970 feet.

341. *Lightning Tubes.*—When lightning descends into a sandy
soil, the sand is sometimes melted by the discharge, and the path
of the lightning is marked by a tube of vitrefied sand.   Such a
tube is called a *fulgurite.*   These tubes are sometimes three inches
in external diameter, with sides nearly an inch in thickness, and
they sometimes extend to a depth of thirty feet.   The inside part
of lightning tubes is smooth and very bright.   It scratches glass,
and strikes fire as a flint.   By passing a powerful electrical discharge through a mixture of sand and salt, similar tubes have
been produced artificially.

342. *Geographical distribution of Thunder-storms.* —Thunder-
storms occur most frequently in the equatorial regions, and diminish as we proceed toward the poles.   From the equator to latitude
30° the average number of thunder-showers annually is 52; from
latitude 30° to latitude 50° it is 20; from latitude 50° to 60° it is
15; and from latitude 60° to 70° it is only 4.   Beyond latitude
70° lightning is of very rare occurrence; and beyond the parallel
of 75° it is believed to be entirely unknown.   •

Within the tropics, where the trade winds prevail, thunder-
storms are rare; but in those calm regions where there is no
steady prevalent wind they are of frequent occurrence.   They
are produced by ascending columns of air in the form of tornadoes; they cover but a small area, commence suddenly, and rarely last over half an hour.

In Lower Peru, where it never rains, thunder is never heard.

Thunder-storms are most frequent in warm climates, because
here evaporation supplies electricity in the greatest abundance,
and the vapor of the air is precipitated most copiously.   In the
middle latitudes thunder occurs chiefly in the summer months,
and it is most frequent about the middle of the afternoon.

343. *Lightning caused by Volcanoes.*—The eruptions of volcanoes are frequently accompanied by vivid flashes of zigzag lightning.   This electricity is probably developed in the same way as
the electricity of common thunder-storms.   The volcano shoots

up to a great height vast volumes of heated air. This air is cooled by elevation, its vapor is condensed, and a cloud is formed. This cloud serves as a conductor for the electricity previously existing in the air, by which means it becomes highly charged, and the electricity thus collected is discharged upon the peak of the volcano.

For the same reason, violent whirlwinds and water-spouts are generally attended by thunder and lightning.

344. *Telegraph Wires affected by Thunder-storms.*—The wires of the electric telegraph present conductors of electricity of vast extent, and they are powerfully affected during the passage of a thunder-storm. The electricity of a distant cloud is sufficient to charge a telegraph wire, and when the electricity of the cloud is discharged, a spark is perceived wherever there is a small interruption in the telegraph wire. This effect is produced at a distance of several miles, and during summer these sparks are often seen in telegraph offices, being sometimes caused by a thunder-storm so remote that no lightning is perceived at the place of observation.

345. *Pointed Objects tipped with Light.*—If in a dark room we hold a pointed conductor near to an electrified body, we may observe the point to be tipped with light. Similar phenomena often occur in nature upon a grand scale. When the lower atmosphere is highly electrified, pointed objects are sometimes seen tipped with light. The tops of the masts, and the ends of the spars of ships, the lances of soldiers, the tips of horses' ears, the point of an umbrella, and similar pointed objects, are frequently luminous at night. Sometimes the hair of the head stands erect, and appears tipped with flame.

All these phenomena are due to a moderate charge of electricity, not sufficient to force its way explosively, but escaping by a gentle current.

SECTION III.
AURORA POLARIS.

346. The aurora polaris is a luminous appearance frequently seen near the horizon as a diffuse light like the morning twilight, whence it has received the name of aurora. In the northern hemisphere it is usually termed *aurora borealis*, because it is chiefly seen in the north. A similar phenomenon is seen in the south-

ern hemisphere, where it is called the *aurora australis*. Each of
them may with greater propriety be called aurora polaris, or *polar light*.

347. *Varieties of Aurora.*—Auroras exhibit an infinite variety
of appearances, but they may generally be referred to one of the
following classes:

First. A horizontal light like the morning aurora or break of·
day. The polar light may generally be distinguished from the
true dawn by its position in· the heavens, since in the United
States it always appears in the northern quarter. This is the
most common form of aurora, but it is not an essentially distinct
variety, being due to a blending of the other varieties in the distance. The upper limit of the light is an arc of a small circle,
which, though indefinite, is better defined than the twilight.

348. *Second. An Arch of Light somewhat in the form of a Rainbow.*—This arch frequently extends entirely across the heavens
from east to west, and cuts the magnetic meridian nearly at right
angles. This arch does not long remain stationary, but frequently rises and falls; and when the aurora exhibits great splendor,
several parallel arches are often seen at the same time, appearing
as broad belts of light, stretching from the eastern to the western
horizon. In the polar regions, five, six, and even seven such
arches have been seen at once; and on two occasions have been
seen *nine* parallel arches separated by distinct intervals. Fig. 68
represents auroral arches seen a few years since in Canada.

Fig. 68.

349. *Third. Slender, luminous beams or columns,* well-defined and often of a bright light. These beams rise to various heights in the heavens from 2° or 3° up to 90° or more; sometimes, though rarely, passing the zenith, Fig. 69. Their breadth varies from a

Fig. 69.

quarter of a degree up to two or three degrees. Frequently they last but a few minutes, sometimes they continue a quarter of an hour, a half hour, or even a whole hour. Sometimes they remain at rest, and sometimes they have a quick lateral motion. This light is commonly of a pale yellow, sometimes reddish, occasionally crimson, or even of blood color. Sometimes the luminous beams are interspersed with dark rays resembling dense smoke. Sometimes the tops of the beams are pointed, and, having a waving motion, they resemble the lambent flames of half-extinguished alcohol burning upon a broad flat surface, Fig. 70, page 176. Faint stars are visible through the substance of the beams.

350. *Fourth. The Corona.*—Luminous beams sometimes shoot up simultaneously from nearly every part of the horizon, and converge to a point a little south of the zenith, forming a quivering canopy of flame, which is called the corona. The sky now resembles a fiery dome, and the crown appears to rest upon variegated fiery pillars, which are frequently traversed by waves or flashes of light. This may be called a *complete aurora,* and comprehends most of the peculiarities of the other varieties. See Fig. 74.

Fig. 70.

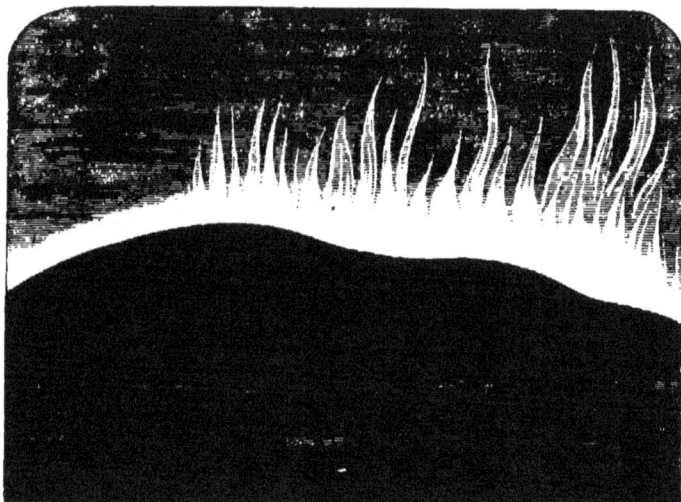

The corona seldom continues complete longer than one hour. The streamers then become fewer and less intensely colored; the luminous arches break up, while a dark segment is still visible near the northern horizon, and at last nothing remains but masses of delicate cirro-cumulus clouds.

During the exhibition of brilliant auroras, delicate fibrous clouds are commonly seen floating in the upper regions of the atmosphere; and on the morning after a great nocturnal display we sometimes recognize the same streaks of cloud which had been luminous during the preceding night. Sometimes during the day these clouds arrange themselves in forms similar to the beams of the aurora, constituting what has been called a *day aurora.*

351. *Fifth. Waves or Flashes of Light.* — The luminous beams sometimes appear to shake with a tremulous motion; flashes like waves of light roll up toward the zenith, and sometimes travel along the line of an auroral arch. Sometimes the beams have a slow lateral motion from east to west, and sometimes from west to east. These sudden flashes of auroral light are known by the name of *merry dancers,* and form an important feature of nearly every splendid aurora.

352. *Duration of Auroras.*—The duration of auroras is very variable. Some last only an hour or two; others last all night; and occasionally they appear on two successive nights under circumstances which lead us to believe that, were it not for the light of the sun, an aurora might be seen uninterruptedly for 36 or 48 hours. For more than a week, commencing August 28th, 1859, in the northern part of the United States, the aurora was seen almost uninterruptedly every clear night. In the neighborhood of Hudson's Bay, the aurora is seen for months almost without cessation.

353. *Recurring Fits.*—Auroras are characterized by recurring fits of brilliancy. After a brilliant aurora has faded away, and almost wholly disappeared, it is common for it to revive, so as to rival and often to surpass its first magnificence. Two such fits are common features of brilliant auroras, and sometimes three or four occur on the same night.

354. *Colors of the Aurora.*—The color of the aurora is very variable. If the aurora be faint, its light is usually white or a pale yellow. When the aurora is brilliant, the sky exhibits at the same time a great variety of tints; some portions of the sky are nearly white, but with a tinge of emerald green; other portions are of a pale yellow, or straw color; others are tinged with a rosy hue; while others have a crimson hue, which sometimes deepens to a blood red. These colors are ever varying in position and intensity.

355. *Geographical Extent of Auroras.*—Auroras are sometimes observed simultaneously over large portions of the globe. The aurora of August 28, 1859, was seen over more than 140 degrees of longitude, from California to Eastern Europe, and from Jamaica, on the south, to an unknown distance in British America, on the north. The aurora of September 2, 1859, was seen at the Sandwich Islands; it was seen throughout the whole of North America and Europe; and the magnetic disturbances indicated its presence throughout all Northern Asia, although the sky was overcast, so that at many places it could not be seen. An aurora was seen at the same time in South America and New Holland.

M

The auroras of September 25, 1841, and November 17, 1848, were almost equally extensive.

356. *Dark Segment.*—In the United States an aurora is uniformly preceded by a hazy or slaty appearance of the sky, particularly in the neighborhood of the northern horizon. When the auroral display commences, this hazy portion of the sky assumes the form of a dark bank or segment of a circle in the north, rising ordinarily to the height of from five to ten degrees, Fig. 71. This

Fig. 71.

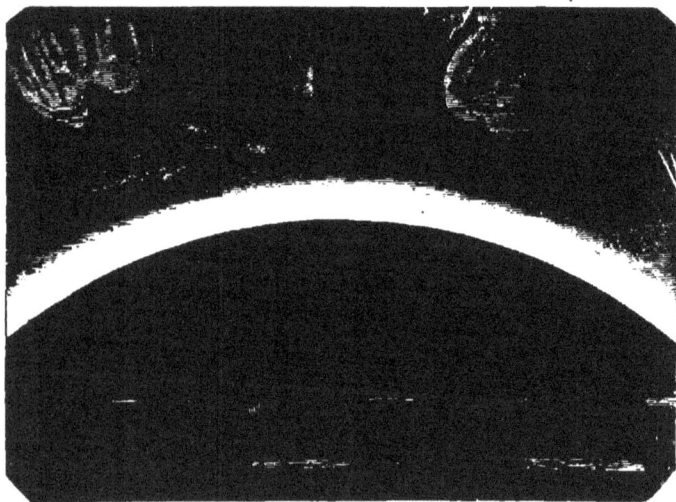

dark segment is not a cloud, for the stars are seen through it, as through a smoky atmosphere with little diminution of brilliancy.

This dark bank is simply a dense haze, and it appears darker from the contrast with the luminous arc which rests upon it. In high northern latitudes, when the aurora covers the entire heavens, the whole sky seems filled with a dense haze; and still nearer the pole, where the aurora is sometimes seen in the south, this dark segment is observed resting on the southern horizon, and bordered by the auroral light. This phenomenon was visible in the United States in the aurora of August, 1859.

The highest point of this dark segment is generally found in the magnetic meridian. Exceptions, however, frequently occur, and in some places there is a constant deviation of ten degrees or more.

· **357.** *Position of Auroral Arches.*—The dark segment is bounded by a luminous arc, whose breadth varies from half a degree to one or two degrees. The lower edge is well defined; but, unless the breadth be very small, the upper edge is ill defined, and blends with a general brightness of the sky. If the aurora becomes brilliant, other arcs usually form at greater elevations, sometimes passing through the zenith.

The summit of these arcs is situated nearly in the magnetic meridian, and the arc sometimes extends symmetrically on each side toward the horizon. Frequently, however, the summit of the arc deviates ten degrees or more from the magnetic meridian, and in some places this deviation appears to be tolerably constant.

Sometimes the arch is incomplete, extending only part of the way from one horizon to the other.

358. *Breadth of Auroral Arches.*—The apparent breadth of auroral arches varies with their elevation above the horizon. The result of a large number of observations made in Scandinavia gave *seven degrees* as the average breadth of arches seen in the north at altitudes less than 60°; for arches seen in the south at altitudes less than 60°, the average breadth was *eight degrees;* while for arches between the limits of 30° zenith distance either north or south, the average breadth was *twenty-five* degrees.

When an arch appears to move across the sky from north to south, or the reverse, its angular breadth exhibits corresponding changes.

If the distance of an arch from the earth remained constant during its movement of translation, and the arch was of the form of a ring whose section was a circle, its breadth when in the zenith should be double that at an elevation of 30°. But its actual breadth in the former case is three or four times as great as in the latter, showing that the greatest breadth of a section of the ring is parallel to the earth.

359. *Form of Auroral Arches.*—Auroral arches are not arcs of great circles; that is, they do not cut the horizon at points 180° from each other. Careful measurements made at several points of some of the most remarkable arcs have shown that, except near the horizon, they may be regarded as portions of small circles parallel to the earth's surface. Such a circle seen obliquely would have the appearance of an ellipse. Near the horizon the elliptic form of the auroral arch has sometimes been quite noticeable, the extremities of the arch being bent inward. Occasionally an ellipse has been seen almost entire, and there is one instance on record in which the ellipse appeared complete, the diameters of the ellipse being as two to one, and the centre of the ellipse being elevated about 15° above the horizon.

360. *Anomalous forms of Arches.*—Sometimes an auroral arch consists of rays arranged in irregular and sinuous bands of various and variable curvatures, presenting the appearance of the undulations of a ribbon or flag waving in the breeze. Sometimes the appearance is that of a brilliant curtain whose folds are agitated by the wind, Fig. 72. These folds sometimes become

Fig. 72.

very numerous and complex, and the arch assumes the form of a long sheet of rays returning into itself, the folds enveloping each other, and presenting an immense variety of the most graceful curves. Sometimes these curves are continually changing, and develop themselves like the folds of a serpent.

361. *Movements of Auroral Arches.*—An auroral arch does not maintain invariably a fixed position. It is frequently displaced, and is transported parallel to itself from north to south, or from south to north. An arch which first appears near the northern horizon sometimes rises gradually, attains the zenith, descends toward the southern horizon, remains there for a time stationary, and then perhaps retraces its course. A series of observations in Scandinavia presented sixty cases in which auroral arches moved from north to south, and thirty-nine cases from south to north. In the United States the motion from north to south is about ten times as frequent as the motion from south to north.

Sometimes there is a movement of the arch from west to east, or from east to west.

The rate of motion of arches is very variable. The angular motion of translation sometimes amounts to 17° per minute, and frequently amounts to 5° per minute. With a vertical elevation of 125 miles above the earth, the last rate of motion would imply an actual velocity of 1000 feet per second.

362. *Structure of Auroral Arches.*—Auroral arches generally tend to divide into short rays running in the direction of the breadth of the arch, and converging toward the magnetic zenith. They frequently seem to be formed of transverse fibres terminating abruptly in a regular curve, which forms the lower edge of the arch, Fig. 73. Arches entirely nebulous are not the most fre-

Fig. 73.

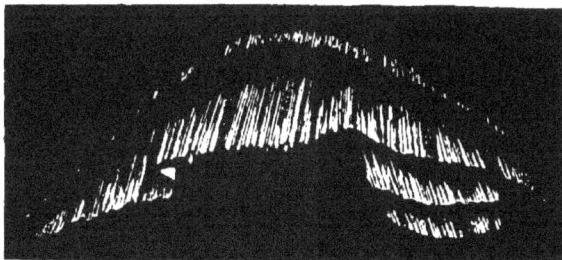

quent; striated arcs are very common, and auroral arches present every intermediate variety between these two extremes. Frequently a nebulous arc resolves itself into a striated arc without changing its general form. Sometimes the rays are distinct and isolated. In such a case the arch generally increases in breadth, extending on the side of the zenith. Sometimes auroral beams arrange themselves in the form of an arch, which is subsequently replaced by an arch of nebulous matter. When the light of the rays is uniform, the dark intervening spaces sometimes present the appearance of *dark rays* or *black striæ* perpendicular to the arch. Sometimes an auroral arch is formed of short streams parallel to each other, presenting the appearance of a row of comet's tails.

363. *Motion of Auroral Beams.*—This motion is either longitudinal, the beam extending toward the zenith or the horizon, or it is a lateral movement which displaces the beam parallel to itself. Frequently a beam extends suddenly either upward or downward. This motion is most common downward, and sometimes with very great velocity. It sometimes takes place simultaneously in a large number of neighboring beams. When a beam rises and falls alternately without any considerable change of length, it is said to *dance*. This is a common occurrence in high latitudes, where it is known by the name of the *merry dancers*.

Beams sometimes move laterally from east to west, and sometimes from west to east; but in the United States the former motion is the most common. Beams advance either from north to south, or from south to north, but the former motion is the most common.

364. *The Corona.*—When the sky is filled with a large number of separate beams all parallel to each other and to the direction of the dipping needle, according to the rules of perspective, these beams will seem to converge to one point, viz., the magnetic zenith, or the point toward which the dipping needle is directed, Fig. 74. Hence results the appearance of a corona, or crown of rays, whose centre is generally, but not always dark.

Numerous measurements have been made of the position of the corona, and they show that the centre of the corona is always very near the magnetic zenith, but not always exactly coincident with it.

The corona is sometimes incomplete, sectors of greater or less extent being deficient. The passage of a striated arch over the magnetic zenith frequently presents the appearance of a corona. If the arch advances from north to south, before reaching the magnetic zenith it forms a half crown on the northern side; at the instant of passing the magnetic zenith we have a complete corona of an elliptic form, whose rays descend nearly to the hori-

Fig. 74.

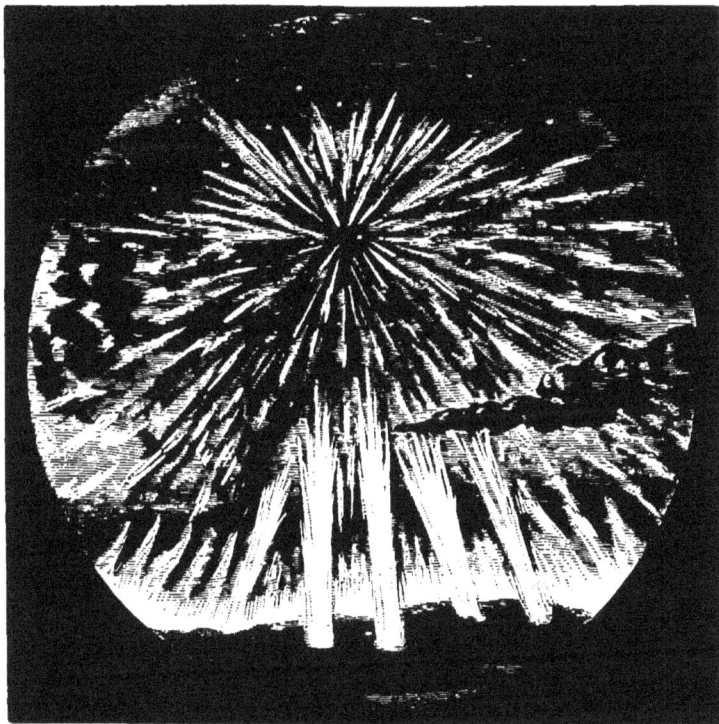

zon on the eastern and western sides; and after the arch has passed the magnetic zenith there is formed a half crown on the southern side.

365. *Auroral Clouds.*—When an aurora becomes less active its beams become less luminous, their edges become more diffuse, they increase in breadth while they diminish in length, and assume the appearance of luminous clouds. Sometimes they exhibit a fibrous structure, and present a strong resemblance to cirrus clouds. These auroral clouds generally make their appearance later in the evening than arches or beams.

366. *Auroral Vapor.*—During the exhibition of a brilliant aurora there is frequently an appearance of general nebulosity or luminous vapor covering large portions of the heavens, and sometimes almost the entire celestial vault. Its light is generally faint, especially in the upper part of the sky, sometimes but little exceeding that of the milky way; but sometimes, near the horizon, the light is intense, resembling a vast conflagration. This seems to indicate that the vertical thickness of the auroral vapor is small in comparison with its horizontal dimensions.

This auroral vapor may appear during any phase of a grand aurora, and is frequently seen during the intervals between the disappearance and reappearance of arches and beams.

367. *Height of the Aurora.*—The great auroral exhibition of August and September, 1859, was very carefully observed at a large number of stations, and these observations afford the materials for determining the height of the aurora above the earth's surface.

At the most southern stations where these auroras were observed, the light rose only a few degrees above the northern horizon; at more northern stations the aurora rose higher in the heavens; at certain stations it just attained the zenith; at stations farther north the aurora covered the entire northern heavens, as well as a portion of the southern; and at places farther north nearly the entire visible heavens from the northern to the southern horizon were overspread with the auroral light.

In Fig. 75, AB represents a portion of the earth's surface, and beneath are given the names of some of the places where observ-

Fig. 75.

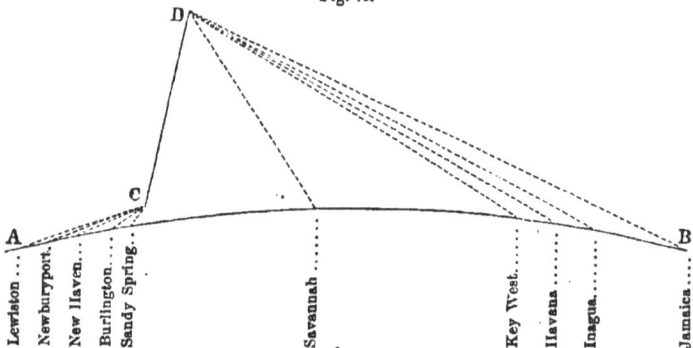

ations were made upon the aurora of August 28, 1859, all at the same hour of the evening. The dotted lines drawn from the five most southern stations represent the elevations of the upper boundary of the auroral light above the northern horizon. The point D thus determined is then the upper edge of the auroral light, near its southern margin, and this point is found to be 534 miles above the earth's surface. The dotted lines from the five most northern stations show the elevation of the lower limit of the auroral light above the south horizon. The point C thus determined is the lower edge of the auroral light, near its southern margin, and this point is found to be 46 miles above the earth's surface. The line CD represents, therefore, the southern boundary of the auroral illumination.

These results, combined with a vast number of other observⁱ ations, show that the aurora of August 28th, 1859, formed a stratum of light encircling the northern hemisphere, extending southward in North America to latitude 38°, and reaching to an unknown distance on the north; and it pervaded more or less the entire interval between the elevations of 46 miles and 500 miles above the earth's surface. This illumination consisted chiefly of luminous beams or columns every where nearly parallel to the di-·rection of a magnetic needle when freely suspended; that is, in the United States the upper extremities of these beams inclined southward at angles varying from 15° to 30°. These beams were therefore about 500 miles in length, and their diameters varied from 5 to 50 miles, and perhaps sometimes they were still greater.

The height of a large number of auroras has been computed by similar methods, and the average result for the upper limit of the streamers is 450 miles.

From a multitude of observations, it is concluded that the aurora seldom appears at an elevation less than about 45 miles above the earth's surface, and that it frequently extends upward to an elevation of 500 miles. Auroral arcs having a well-defined border are generally less than 100 miles in height.

368. *Conflicting Estimates of the Height.*—Some persons contend that the aurora is sometimes seen at elevations of less than one mile above the earth's surface. It is claimed that the aurora is sometimes seen between the observer and a cloud, but this appearance is believed to result from a cloud of very small density strongly illumined by auroral light, which shines through the cloud so as to produce the same appearance as if the aurora prevailed on the under side of the cloud.

Sometimes the lower extremity of an auroral streamer appears to be prolonged below the summit of a neighboring mountain or hill. This appearance is probably an illusion. The same phenomenon has been noticed by careful observers, who ascribed the result to the reflection of the auroral light from the snow which covered the mountains.

Although it is possible that the aurora may sometimes descend nearly to the earth's surface, there is no sufficient evidence to prove that the true polar light has ever descended so low as the region of ordinary clouds.

369. _Noise of the Aurora._ — There is no satisfactory evidence that the aurora ever emits any audible sound. It is a common impression, at least in high latitudes, that the aurora sometimes emits sound. This sound has been called a rustling, hissing, crackling noise. But the most competent observers, who have spent several winters in the Arctic regions, where auroras are seen in their greatest brilliancy, have been convinced that this supposed rustling is a mere illusion. It is therefore inferred that the sounds which have been ascribed to the aurora must have been due to other causes, such as the motion of the wind, or the cracking of the snow and ice in consequence of their low temperature.

If the aurora emitted an audible sound, this sound ought to follow the auroral movements after a long interval. Sound requires four minutes to travel fifty miles. But the observers who report auroral noises make no mention of any interval. It is therefore inferred that the sounds which have been heard during auroral exhibitions are to be ascribed to other causes than the aurora.

370. _Geographical Distribution of Auroras._ — Auroras are very unequally distributed over the earth's surface. They occur most frequently in the higher latitudes, and are almost unknown within the tropics. At Havana, latitude 23°, but six auroras have been recorded within a hundred years, and south of Havana auroras are still more unfrequent. As we travel northward from Cuba, auroras increase in frequency and brilliancy; they rise higher in the heavens, and oftener attain the zenith. Near the parallel of 40°, we find, on an average, only ten auroras annually. Near the parallel of 42°, the average number is twenty annually; near 45°, the number is forty; and near the parallel of 50°, it amounts to eighty annually. Between this point and the parallel of 62°, auroras are seen almost every night. They appear high in the heavens, and as often to the south as the north. Farther north they are seldom seen except in the south, and from this point they diminish in frequency and brilliancy as we advance toward the pole. Beyond latitude 62° the average number of auroras is reduced to forty annually. Beyond latitude 67° it is reduced to twenty, and near latitude 78° to ten annually.

Fig. 76.

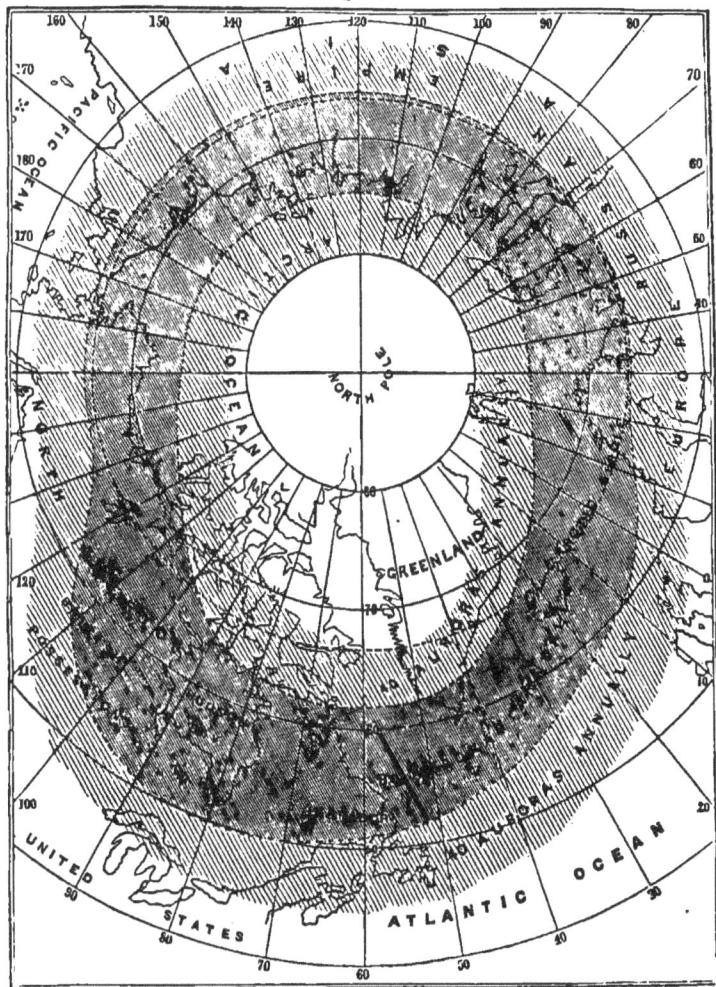

If we make a like comparison for the meridian of St. Petersburg, we shall find a similar result, except that the auroral region is situated farther northward than it is in America, the region of eighty auroras annually being found between the parallels of 66° and 75°.

Upon Fig. 76, the dark shade indicates the region where the average number of auroras annually amounts to at least eighty, and the lighter shade indicates the region where the average number of auroras annually amounts to at least forty.

We thus see that the region of greatest auroral action is a zone of an .oval form surrounding the north pole, and whose central line crosses the meridian of Washington in latitude 56°, and the meridian of St. Petersburg in latitude 71°. Accordingly, auroras are much more frequent in the United States than they are in the same latitudes of Europe.

The form of this auroral zone bears considerable resemblance to a magnetic parallel, or line every where perpendicular to a magnetic meridian, and it is probable that there is a real connection between the two phenomena.

371. *Auroras in the Southern Hemisphere.*—Auroras in the southern hemisphere are nearly, if not quite as frequent as they are in the corresponding magnetic latitudes of the northern hemisphere, and it is probable that the geographical distribution of auroras in the two hemispheres is somewhat similar.

372. *Simultaneous Auroras in both Hemispheres.*—By comparing the records of auroras in the two hemispheres we find a remarkable coincidence of dates, which seems to justify the conclusion that an unusual auroral display in the southern hemisphere is always accompanied by an unusual display in the northern hemisphere; that is, a great exhibition of auroral light about one magnetic pole of the earth, is uniformly attended by a great exhibition of auroral light about the opposite magnetic pole.

373. *Diurnal Periodicity of Auroras.*—Auroras appear at all hours of the night, but not with equal frequency. The average number increases uninterruptedly from sunset till about midnight, from which time the number diminishes uninterruptedly till morning. In Canada the maximum occurs an hour before midnight; farther north, in latitude 52°, the maximum occurs at midnight; and still farther north to the Arctic Ocean, the maximum occurs an hour after midnight.

374. *Annual Periodicity of Auroras.*—Auroras occur in each month of the year, but not with equal frequency. In New England and New York the least number of auroras is recorded in winter, and the greatest number in the autumn; but if we make allowance for the diminished length of the nights in summer, we must conclude that auroras are about as frequent in summer as in autumn. There is a decided diminution in the frequency of auroras in winter, and a period of maximum frequency from April to September, with perhaps a slight diminution during the intervening month of June.

Observations in Canada lead to similar conclusions, except that the unequal length of the days has a somewhat greater influence upon the number of auroras recorded in summer.

375. *Secular Periodicity of Auroras.*—The number of auroras seen in different years is extremely variable. Sometimes for several years auroras are remarkable for their number and magnificence, and then succeeds a barren interval during which auroras are almost entirely forgotten.

If we compare the observations made at any one station for a long period of years, we shall discover that the inequality in the number of auroras upon successive years bears a strong resemblance to a secular periodicity.

From a continued series of observations of the aurora made at Boston and New Haven from 1742 to the present time, it appears that auroras were unusually frequent from 1780 to 1791, and again from 1837 to 1854; but from 1742 to 1768, and again from 1792 to 1837, they were much less frequent and brilliant. These observations indicate two periods of maximum abundance, the culminating points of which were about 1787 and 1845; that is, during the past century the frequency of auroras in New England has been subject to an inequality bearing considerable resemblance to an astronomical periodicity, the period being about fifty-eight years.

A comparison of European observations for the past two centuries leads to similar conclusions. We find periods of unusual abundance culminating about 1728, 1780, and 1842; and between these periods we find long intervals of great barrenness. These results show a considerable degree of uniformity approximating toward a period of fifty-nine years from one maximum to another.

At the same time, the observations indicate important exceptions, which seem to point to a subordinate period of ten years. In both the American and European observations there is a frequent alternation of meagre and abundant years, and the intervals do not differ much from ten years. The observations then seem to indicate a maximum every ten years, and a maximum of extraordinary brilliancy every fifty-nine or perhaps sixty years.

376. *Disturbance of the Magnetic Needle.*—The aurora is ordinarily accompanied by a considerable disturbance of the magnetic needle, and the effect increases with the brilliancy and extent of the aurora. Auroral beams cause a disturbance of the needle, particularly when the beams themselves are in active motion. Auroral waves or flashes, especially if they extend to the zenith, cause a violent agitation of the needle, consisting of an irregular oscillation on each side of its mean position.

These extraordinary deflections of the needle prevail almost simultaneously over large portions of the globe, even where the aurora itself is not visible. During the great auroral display of September 2, 1859, the disturbances of the magnetic needle were very remarkable throughout North America, Europe, and Northern Asia, as well as in New Holland. At Toronto the declination of the needle changed 3° 45′ in half an hour. The inclination was observed to change 2° 49′ when the needle passed beyond the limits of the scale, so that the entire range of the needle could not be determined. The horizontal force was observed to change to the extent of *one ninth* of its whole value when the needle passed beyond the limits of the scale, so that its entire range was unknown. At several observatories in Europe still more remarkable disturbances were recorded.

377. *Progress of Magnetic Disturbances.*—These irregular deflections of the magnetic needle are not quite simultaneous at distant stations. Over the surface of Europe they appear to be propagated in a direction from N. 28° E. to S. 28° W. at the rate of about 100 miles per minute. Over the surface of North America they are propagated at about the same velocity in a direction from N. 68° E. to S. 68° W.

378. *Influence of the Aurora upon the Telegraph Wires.*—Auroras exert a remarkable influence upon the wires of the electric telegraph. During the prevalence of brilliant auroras the telegraph lines generally become unmanageable. The aurora develops electric currents upon the wires, and hence results a motion of the telegraph instruments similar to that which is employed in telegraphing, and this movement, being frequent and irregular, ordinarily renders it impossible to transmit intelligible signals. During several remarkable auroras, however, the currents of electricity on the telegraph wires have been so steady and powerful that they have been used for telegraph purposes as a substitute for a voltaic battery; that is, telegraph messages have been transmitted from the auroral influence alone. This result proves that the aurora develops on the telegraph wires an electric current similar to that of a voltaic battery, and differing only in its variable intensity.

THEORY OF THE POLAR LIGHT.

379. *Is the Aurora caused by Nebulous Matter falling into our Atmosphere?*—Some have ascribed the polar light to a rare nebulous matter occupying the interplanetary spaces, and revolving round the sun at such a distance that a portion of this matter occasionally falls into the upper regions of the atmosphere with a velocity sufficient to render it luminous from the condensation of the air before it. But we can see no reason why matter, reaching the earth from such a source, should always be confined to certain districts of the earth, and be wholly unknown in other portions. This hypothesis, therefore, can not be reconciled with the known geographical distribution of auroras.

380. *Auroral Exhibitions are Terrestrial Phenomena.*—Auroral exhibitions take place in the upper regions of the atmosphere, and partake of the earth's rotation. All the celestial bodies have an apparent motion from east to west, arising from the rotation of the earth; but bodies belonging to the earth, including the atmosphere and the clouds which float in it, partake of this rotation, so that their relative position is not affected by it. The same is true of the aurora. Whenever a corona is formed, it maintains sensibly the same position in the heavens during the whole period of its continuance, although the stars meanwhile revolve at the rate of 15° per hour.

381. *The Auroral Light is Electric Light.* — This is proved by the effect of an aurora upon the telegraph wires. The electric telegraph is worked by a current of electricity generated by a voltaic battery, and flowing along the conducting wire which unites the distant stations. This current, flowing round an electro-magnet, renders it temporarily magnetic, so that its armature is attracted, and a mark is made upon a roll of paper. During a thunder-storm the electricity of the atmosphere affects the conducting wire in a similar manner, and a great auroral display produces a like effect. During the auroras of August and September, 1859, there were remarked all those classes of effects which are considered as characteristic of electricity.

A. In passing from one conductor to another, electricity exhibits a *spark* of light. During the auroras of 1859, at numerous stations both in America and Europe, brilliant sparks were drawn from the telegraph wires when no battery was attached.

B. In passing through poor conductors, electricity develops *heat*. During the auroras of 1859, both in America and Europe, paper and even wood were set on fire by the auroral influence alone.

C. When passed through the animal system, electricity communicates a well-known characteristic *shock*. During the auroras of 1859 several telegraph operators received severe shocks when they touched the telegraph wires.

D. A current of electricity develops *magnetism* in ferruginous bodies. The auroras of 1859 developed magnetism so abundantly and so steadily that it was more than sufficient for the ordinary business of telegraphing.

E. A current of electricity deflects a magnetic needle from its normal position. In England the usual telegraph signal is made by a magnetic needle surrounded by a coil of copper wire, so that the needle is deflected by an electric current flowing through the wire. Similar deflections were caused by the auroras of 1859, and these deflections were greater than those produced by the telegraph batteries.

F. A current of electricity produces chemical *decompositions*. The auroras of 1859 produced the same marks upon chemical paper as are produced by an ordinary voltaic battery; that is, they decomposed a chemical compound.

G. Certain bodies, such as a solution of sulphate of quinine, when illumined by an electric spark, present a very peculiar appearance, as if they were self-luminous. This appearance is termed *fluorescence*. The same effect is produced upon these substances by the auroral light.

These facts demonstrate that the fluid developed by the aurora on telegraph wires is indeed electricity. This electricity may be derived from the aurora either by transfer or by induction. If we adopt the former supposition, then the auroral light is certainly electric light. If we adopt the latter supposition, then, since we know of but two agents, magnetism and electricity, capable of inducing electricity in a distant conductor, and since the auroral fluid is luminous while magnetism is not luminous, we must admit that the auroral light is electric light.

382. *Colors of the Aurora.*—The colors of the aurora are the same as those of ordinary electricity passed through rarefied air. When a spark is drawn from an ordinary electrical machine in air of the usual density, the light is intense and nearly white. If the electricity be passed through a glass vessel in which the air has been partially rarefied, the light is more diffuse, and inclines to a delicate rosy hue. If the air be still farther rarefied, the light becomes very diffuse, and its color becomes a deep rose or purple. The same variety of colors is observed during the aurora. The transition from a white or pale straw color to a rosy hue, and finally to a deep red, probably depends upon the height above the earth, and upon the amount of condensed vapor present in the air.

The emerald green light which is seen in some auroras is ascribed to the projection of the yellow light of the aurora upon the blue sky, since a combination of yellow and blue light produces green. A similar effect is often produced in the evening twilight by a combination of the yellow light of the sun with the blue of the celestial vault.

383. *The Auroral Corona.*—The formation of an auroral corona near the magnetic zenith is the effect of perspective, resulting

N

from a great number of luminous beams all parallel to each other. A collection of beams parallel to the direction of the dipping needle would all appear to converge toward the pole of the needle, and no other supposition will explain all the appearances which we observe. The auroral crown, therefore, every where appears in the magnetic zenith, and it is not the same crown which is seen at different places any more than it is the same rainbow which is seen by different observers.

384. *What are Auroral Beams?*—The auroral beams are simply illumined spaces caused by the flow of electricity through the upper regions of the atmosphere. During the auroras of 1859 these beams were nearly 500 miles in length, and their lower extremities were elevated about 45 miles above the earth's surface. Their tops inclined toward the south about 17° in the neighborhood of New York.

It was formerly supposed that the electric current necessarily moved in the direction of the axis of the auroral beams; that is, that the electric discharge was from the upper regions of the atmosphere to the earth, or the reverse. Recent discoveries throw some doubt upon this conclusion. When a stream of electricity flows through a vessel from which the air is almost wholly exhausted, under certain circumstances the light becomes stratified, exhibiting alternately bright and dark bands crossing the electric current at right angles, from which it might be inferred that electricity flowing horizontally through the upper regions of the atmosphere might exhibit alternately bright and dark bands like the auroral beams. But this stratification of the electric light is due to *intermittences* in the intensity of the electric discharge, and it is not probable that such intermittences can take place in nature with sufficient rapidity to produce a similar effect. It is therefore more probable that auroral beams are the result of a current of electricity traveling in the direction of the axis of the beams.

385. *Cause of the Dark Segment.*—The slaty appearance of the sky which is remarked in all great auroral exhibitions arises from the condensation of the vapor of the air, and this condensed vapor probably exists in the form of minute spiculæ of ice or flakes of snow. Fine flakes of snow have been repeatedly observed to fall during the exhibition of auroras, and this snow only slightly impairs the transparency of the atmosphere, without presenting the appearance of clouds. It produces a turbid appearance of the atmosphere, and causes that dark bank which in the United States rests on the northern horizon. This turbidness is more noticeable near the horizon than it is at great elevations, because near the horizon the line of vision traverses a greater depth of this hazy atmosphere. When the aurora covers the whole heavens, the entire atmosphere is filled with this haze, and a dark segment may be observed resting on the southern horizon.

386. *Circulation of Electricity about the Earth.*—The vapor which rises from the ocean in all latitudes, but most abundantly in the equatorial regions of the earth, carries into the upper regions of the atmosphere a considerable quantity of positive electricity, while the negative electricity remains in the earth. This positive electricity, after rising nearly vertically with the ascending currents of the atmosphere, would be conveyed toward either pole by the upper currents of the atmosphere.

The earth and the rarefied air of the upper atmosphere may be regarded as forming the two conducting plates of a condenser, which are separated by an insulating stratum, viz., the lower portion of the atmosphere. The two opposite electricities must then be condensed by their mutual influence, especially in the polar regions, where they approach nearest together, and whenever their tension reaches a certain limit, there will be discharges from one conductor to the other. When the air is humid it becomes a partial conductor, and conveys a portion of the electricity of the atmosphere to the earth. On account of the low conducting power of the air, the neutralization of the opposite electricities would not be effected instantaneously, but by successive discharges, more or less continuous, and variable in intensity. These discharges should frequently occur simultaneously at the two poles, since the electric tension of the earth should be nearly the same at each pole.

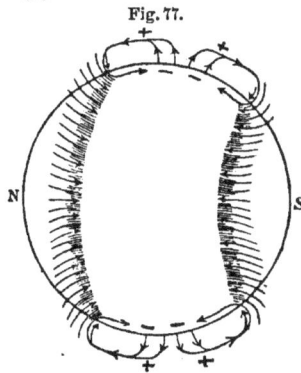

Fig. 77.

Fig. 77 represents the system of circulation here supposed, the north and south poles of the earth being denoted by the letters N. and S.

387. *Cause of the Auroral Beams.*—When electricity from the upper regions of the atmosphere discharges itself to the earth through an imperfectly conducting medium, the flow can not be every where uniform, but must take place chiefly along certain lines where the resistance is least; and this current must develop light, forming thus an auroral beam. It might be expected that these beams would have a vertical position, but their position is controlled by the earth's magnetism. It is found that when magnetic forces act upon a perfectly flexible conductor through which an electric current passes, the conductor must assume the form of a magnetic curve. Now at each point of the earth's surface the dipping needle shows the direction of the magnetic curve passing through that point. Hence the axis of an auroral streamer must lie in the magnetic curve which passes through its base; and since adjacent streamers are sensibly parallel, the beams appear to converge toward the magnetic zenith.

388. *Position of Auroral Arches explained.*—When electricity escapes from a metallic conductor under a receiver from which the air has been exhausted, and this conductor is the pole of a powerful magnet, the electric light forms a complete luminous ring around this conductor.

In like manner, the auroral arch is a part of a luminous ring nearly parallel to the earth's surface, having the magnetic pole for its centre, and cutting all the magnetic meridians at right angles; and this position results from the influence of the earth's magnetism.

389. *Anomalous Position of Auroral Arches.*—Auroral arches are not always exactly perpendicular to the magnetic meridian, and in some places this deviation is uniform, and may amount to ten degrees. Such a deviation may be explained as follows:

The direction of the magnetic needle at any place is determined mainly by its position with respect to the magnetic poles of the earth, but partly by local causes, such as the conformation of the land and sea, etc. In consequence of these local causes, the direction of the magnetic needle at some places probably differs several degrees from what it would be if it were controlled entirely by the magnetic poles. This local influence probably diminishes as we rise above the earth's surface, so that at the height of the auroral streamers the direction of the magnetic needle may differ several degrees from that at the surface of the earth.

390. *Cause of the Auroral Flashes.*—The flashes of light observed in great auroral displays are 'due to *inequalities* in the motion of the electric currents. On account of the imperfect conducting power of the air, the flow of electricity is not perfectly · uniform, but escapes by paroxysms. The flashes of the aurora are therefore feeble flashes of lightning.

391. *Cause of the Magnetic Disturbances.*—The disturbance of the magnetic needle during auroras is due to currents of electricity flowing through the atmosphere or through the earth. A magnetic needle is deflected from its mean position by an electric current flowing near it through a good conductor like a copper wire. A stream of electricity flowing through the earth or the atmosphere must produce a similar effect.

It is probable that the directive power of the magnetic needle is due to electric currents circulating around the globe from east to west. Such currents would cause the magnetic needle every where to assume a position corresponding with what is actually observed; and the existence of such currents has been proved by direct observation.

According to Art. 386, positive electricity circulates from the equator toward either pole through the upper regions of the atmosphere, and thence through the earth toward the equator, to restore the equilibrium which is continually disturbed by evaporation from the waters of the equatorial seas. This current from the polar regions must modify the regular current which is supposed to be constantly circulating from east to west, resulting in a current from northeast to southwest, in conformity with observations.

This current does not, however, flow uninterruptedly from N.E. to S.W., but alternates at short intervals with a current in the contrary direction. Such currents of electricity must produce a continual disturbance of the magnetic needle, and they are sufficient to account for the disturbances actually observed.

392. *Effect of the Aurora upon Telegraph Wires.*—The effect of the aurora upon the telegraph wires is similar to that of electricity in thunder-storms, except in the intensity and steadiness of its action. During thunder-storms the electricity of the wires is discharged instantly with a flash of lightning, while during auroras there is sometimes a strong and steady flow continuing for several minutes.

393. *Cause of the Diurnal Inequality of Auroras.*—The diurnal inequality in the frequency of auroras is due to the same cause as the diurnal variation in the intensity of atmospheric electricity. The same causes which favor the escape of electricity from the upper atmosphere to the earth will produce an aurora whenever the electricity of the upper air is sufficiently intense, and the conducting power of the air is favorable for the slow transmission of an electric current.

394. *Cause of the Annual Inequality of Auroras.*—The unequal frequency of auroras in the different months of the year depends partly upon the amount of electricity present in the upper air, and partly upon the humidity of the air by which this electricity may be discharged. The supply of electricity must be greatest when the evaporation is most rapid, that is, in summer, and this is probably the reason why in North America auroras are more frequent in summer than in winter. In Europe auroras are seldom seen in midsummer, because in those latitudes where auroras are most frequent, twilight in midsummer continues all night.

395. *Cause of the Secular Inequality of Auroras.*—The secular inequality in the frequency of auroras indicates the influence of distant celestial bodies upon the electricity of our globe. The periods of auroras observe laws which are similar to those of two other phenomena, viz., the mean *diurnal variation of the magnetic needle,* and the frequency of *black spots upon the sun's surface.*

The magnetic needle has a small diurnal variation, the north end moving a little to the east in the morning, and toward the west about the middle of the day. The mean daily change of the needle not only varies with the locality, but also varies from one year to another at the same locality, and these variations present a decided appearance of periodicity. At Prague the mean daily change of the needle in 1838 was 12', from which time the range diminished steadily to 1844, when it was only 6', from which time it increased to 1848, when it amounted to 11', the interval from one maximum to another being a little more than ten years.

Observations made at other places, and extending back nearly a century, indicate a maximum in the range of the magnetic needle every ten or eleven years, but the successive maxima are not equal to each other. They exhibit variations which indicate a periodicity, the greatest values occurring at intervals of from fifty to sixty years. See Table XXXIV.

The relative frequency of the solar spots exhibits a similar periodicity, and the maximum number of spots corresponds with the maximum value of the magnetic variation.

These three phenomena, the solar spots, the mean daily range of the magnetic needle, and the frequency of auroras, exhibit two distinct periods; one a period of from ten to twelve years, the other a period of from fifty-eight to sixty years. The first of these periods corresponds to one revolution of Jupiter, and the other period corresponds to five revolutions of Jupiter, or two of Saturn, and we can scarcely doubt that the above-mentioned phenomena depend upon the movements of these planets. Observations have also indicated subordinate fluctuations which are probably due to the action of Venus.

We do not know how the planets exert an influence upon the sun's surface; but we may suppose that there are circulating round the sun powerful electric currents, which may possibly be the source of the sun's light; these currents may act upon the planets, developing in them electric currents; and the currents circulating round the planets may react upon the solar currents with a force varying with their distances and relative positions, exhibiting periods corresponding to the times of revolution of the planets. These disturbances of the solar currents may be one cause of the solar spots, and an unusual disturbance of the solar currents may cause a disturbance of the currents of the earth's surface, giving rise to unusual displays of the aurora.

396. *Geographical Distribution of Auroras.*—The geographical distribution of auroras depends chiefly upon the relative intensity of the earth's magnetism in different latitudes. According to experiments with artificial magnets, the electric light tends to form a ring around the pole, and at some distance from it. The electric light should therefore be most noticeable in the neighborhood of the earth's magnetic pole, but not directly over it.' Auroras are, accordingly, most abundant along a certain zone which follows nearly a magnetic parallel, being every where nearly at right angles to the magnetic meridian of the place.

397. *Why Auroras do not occur within the Tropics.*—The electricity of the tropical regions has great intensity, and moves with explosive violence in thunder-showers, while the magnetic intensity in those regions is very feeble, and is insufficient to control the movements of the electricity. In the higher latitudes thunder-showers become infrequent, the electricity of the atmosphere passes to the earth in a slow and quiet manner, and these discharges are controlled by the magnetism of the earth.

398. *Cause of the simultaneous Displays in both Hemispheres.*—We can not explain the great auroral displays in the northern hemisphere by supposing that the electricity of the atmosphere is temporarily·diverted from one hemisphere to the other, for the mean range of the magnetic needle exhibits its maxima simultaneously in both hemispheres; neither can we suppose that the absolute amount of electricity for the entire globe, as developed by evaporation from the water of the ocean, should undergo great periodical variations, for the mean temperature of the earth's surface does not change sensibly from one year to another. We seem, therefore, compelled to ascribe these great auroral displays in no small degree to the direct action of the sun through the agency perhaps of its magnetism, or of the electric currents circulating around it. Such an effect should take place simultaneously in both hemispheres.

· **399.** *Possible System of Electrical Circulation.*—Hence it appears probable that great auroral displays are not exclusively atmospheric phenomena, but are to some extent the result of the influence of extra terrestrial forces. But, if these extraordinary electrical currents are mainly determined by celestial forces, then, since the earth exhibits many of the properties of a permanent magnet, the two magnetic poles of the earth ought to exert opposite influences, and we should expect that the currents in the neighborhood of the two poles would move in contrary directions. Hence we naturally infer a system of circulation similar

Fig. 78.

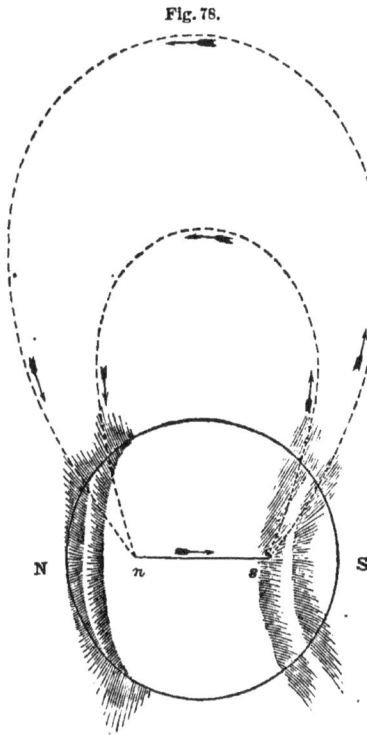

to that which is represented by Fig. 78, where N and S are supposed to represent the north and south magnetic poles of the earth, n and s the poles of an imaginary magnet representing the magnetism of the earth. The east and west bands represent auroral arches upon which stand auroral streamers. The dotted lines represent magnetic curves passing from auroral streamers in the southern hemisphere to streamers in the northern hemisphere, showing the path pursued by the currents of electricity in passing from one hemisphere to the other, above the atmosphere.

This hypothesis agrees substantially with that stated in Art. 386, so far as the phenomena can be observed in the northern hemisphere, but they lead to different results in the southern hemisphere. We have not the requisite observations from the southern hemisphere to enable us to decide between these two hypotheses.

CHAPTER VIII.

OPTICAL METEOROLOGY.

SECTION I.

MIRAGE.

400. Mirage is an atmospheric phenomenon which produces an apparent displacement of distant objects, sometimes elevating and sometimes depressing them; sometimes leaving the image erect, and sometimes inverting it, as when objects are seen reflected

from a lake of tranquil water. It is frequently observed on sandy plains intensely heated by the sun, especially in Egypt and Arabia.

Lower Egypt is a vast sand,plain, with occasional villages situated upon small eminences. In the middle of the day, these villages, seen from a distance, appear as if situated in the midst of a lake, in which are seen the inverted images of houses and trees. The outline of these images is a little indistinct, often exhibiting an undulatory motion, as if reflected from agitated water. As the spectator approaches the boundary of the apparent lake, the waters seem to retire, and the same illusion appears around the next village. Similar phenomena are common in some parts of California, and are occasionally seen in all parts of the United States. Fig. 79, p. 203, represents the mirage as seen in Abyssinia.

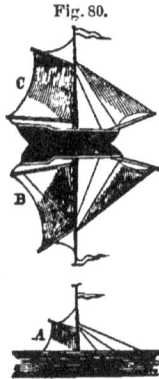

Fig. 80.

Sometimes at sea, when a ship is barely visible in the distant horizon, we perceive above the ship, A, Fig. 80, its inverted image, B, and perhaps above that again a second erect image, C. Sometimes of the two upper images only the inverted one is seen, and sometimes only the erect one.

All these phenomena are due to unusual variations in the refractive power of the air, arising from extreme changes of temperature. The mirage is chiefly seen over a large sandy plain, or over water.

401. Mirage upon a Desert.—Imagine a sandy plain nearly horizontal, and intensely heated by the rays of the sun. The stratum of air which rests upon the sand becomes heated by it; this heat is partially communicated to the superincumbent strata, so that the density of the air increases rapidly as we rise above the earth up to a moderate height.

Let AB, Fig. 81, represent a tree which may be viewed from C in its true position through air of nearly uniform density; and suppose the air beneath it to consist of strata of variable density, decreasing from A to the surface of the ground. The rays of light, AD, BE, which proceed from the top and bottom of the tree, passing successively through strata of less density, will be deviated more and more from a vertical direction, until at last

Fig. 79.

they meet a stratum at such an angle that they are unable to enter it, and they are totally reflected from this stratum at D and

Fig. 81.

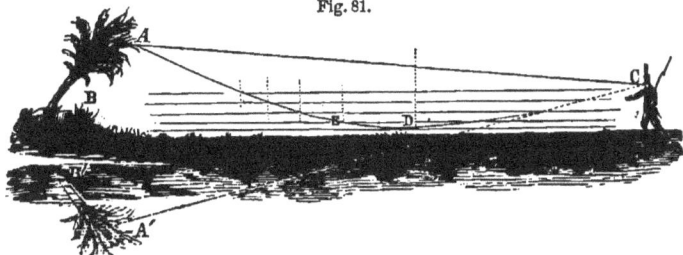

E.   After reflection, these rays, traversing strata more and more dense, will be refracted upward, and at C reach the eye of the observer, who perceives the tree in the direction of the last refracted rays.   An image, A'B', will therefore be seen below the real object, and it will appear inverted, because the rays have suffered reflection.   The effect is similar to that produced by the reflection of a tree from the surface of a tranquil lake, and the observer is thus led to imagine himself to be surrounded entirely by water.

Since the difference of refraction of the successive strata of air is necessarily small, the ray AD must be very oblique; that is, the object must be elevated but little above the ground, and the observer must be at a considerable distance.

402. *Experimental Illustration.*—The mirage may be imitated artificially by superposing in the same vessel two liquids of different densities, such as water and alcohol, or water and sirup of sugar, or simply cold and warm water. These liquids, by partial mixture, produce a medium whose refractive power decreases from the alcohol to the water, so that by looking through this mixture at an object held behind the vessel, an inverted image of it may be seen.

When a sandy plain is intensely heated by the sun, and the air is very calm, if we place the eye near the ground we may generally see the inverted image of grass and other objects at a distance.

403. *Mirage at Sea.*—Mirage is produced at sea when the atmosphere is perfectly calm, and the air in contact with the water is colder and consequently denser than the stratum of air immediately above it; this second stratum is denser than the one next above it, and so on. In such a case, an inverted image of a distant object, as a ship, may be seen with a distinctness almost equal to that of the object itself, and this image will be formed *above* the object.

Let AB, Fig. 82, represent a ship near the horizon seen in

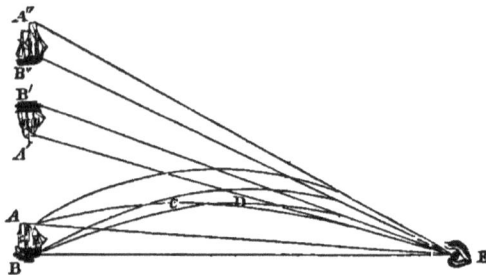

Fig. 82.

its true position by direct rays coming to the eye at E, through strata of air of nearly uniform density. Suppose the air consists of strata of variable density, the density diminishing rapidly from below upward. The rays of light, AD, BC, which proceed from the top and bottom of the ship, passing from a denser to a rarer medium, will be deviated more and more from a vertical, until at last they meet a stratum so obliquely that they are unable to enter it, and are totally reflected from this stratum at D and C. These rays, in passing from the rarer to the denser medium, are now refracted downward, and meet the eye at E, which perceives the ship in the direction of the last refracted rays, and the object appears inverted because the rays have suffered reflection.

Other rays, that never could reach the eye at E in the ordinary state of the atmosphere, may likewise be bent into curves which do not cross before reaching the eye. In this case an erect image of the ship may be seen, and both the direct and inverted images may be seen simultaneously.

404. *Lateral Mirage.*—In mountainous countries, or near a high coast, it may happen that the air is divided by a nearly vertical plane into two portions, one of which is heated by the sun, while the other is in the shadow of a hill or a bank. The transition from the warm to the cold air will not be abrupt, but the density of the vertical sections will increase gradually from the warmer to the colder portion. If an ob-

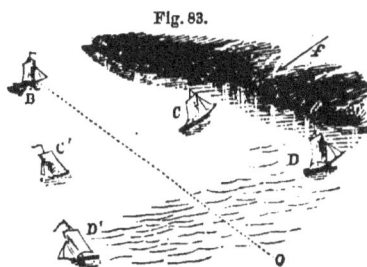

Fig. 83.

server were situated at B, Fig. 83, near this bounding plane, he might see in the warmer part a symmetrical image, C'D', of objects, CD, situated in the colder part, as if in a vertical mirror. This is called a lateral mirage. It is less frequently seen than the other varieties, and its duration is more transient.

405. *Displacements.*—Under certain circumstances, objects near the horizon may appear displaced; sometimes laterally, as in the vicinity of mountains, but more frequently in a vertical direction, in which case they appear elevated above their true position. Sometimes an object appears double, certain rays reaching the eye without sensible deviation, while others, traversing strata of increasing density, describe a curve line. This phenomenon differs from the true mirage in this respect, that the image is not inverted, showing that the light has not suffered reflection.

SECTION II.

ABSORPTION AND REFLECTION OF LIGHT BY THE ATMOSPHERE.

406. *Absorption of Light.*—The atmosphere is never perfectly transparent, but absorbs a portion of the light which traverses it. Hence distant objects, as the summits of mountains, generally appear dim, as if enveloped in a mist or a bluish smoke. This loss of light is due partly to the presence of minute particles of condensed vapor, and also small particles of dust, and partly to the difference of density of the strata arising either from a difference of pressure or a difference of temperature. In passing from one stratum to another of a different density, a portion of light is reflected, so that the transmitted portion is continually diminished. After a rain, when by a general mingling of the strata the temperature of the air has been rendered nearly uniform, its transparency is greatly increased.

407. *Redness of the Evening Sky.*—The redness of the evening sky is due principally to the condensed vapor of the air, a portion of which begins to be precipitated as the temperature of the day declines.

If we transmit the sun's light through a glass prism at different hours of the day, we shall find that the spectrum changes with the altitude of the sun. As the sun approaches the horizon, the violet part of the spectrum contracts, and at length disappears altogether, while the red end of the spectrum remains entire. We hence conclude that the violet rays, which are the most refrangible, have the least power of penetrating the dense atmosphere, including the dust and the condensed vapor near the horizon, and therefore, when the sun is near the horizon, his light exhibits an excess of those rays bordering upon the red end of the spectrum, and this color is communicated not only to the evening sky, but also to the clouds which float in the atmosphere.

From the same cause, the sun, just before setting, sometimes assumes a deep red color, as if seen through a smoked glass, and this redness is more noticeable in the setting than in the rising sun, because in the morning the condensed particles of vapor have descended to the earth, or are converted again into invisible vapor by the increasing heat of the morning.

408. *Reflected Light of the Sky.*—An observer at night, in the neighborhood of a large city, may notice a decided illumination of the heavens, arising from the light of the city reflected from the sky, and during an extensive conflagration this illumination is sometimes very brilliant. The atmosphere therefore reflects a portion of the light which. falls upon it. It is this light of the .sky which prevents our seeing the stars in the daytime, and its brightness is but little inferior to that of the moon, for during the day the moon appears like a small white cloud. It is chiefly from this source that, during the day, apartments which are not accessible to the direct rays of the sun derive their illumination.

The brightness of the sky is variable. It depends upon the purity of the air, increasing with the number of the particles of condensed vapor suspended in it. It depends also upon the weight of the air above the observer, being less on the summits of mountains than at the level of the sea.

The light of the sky is greatest in the vicinity of the sun, and diminishes rapidly as we recede from his disc.

409. *Blue Color of the Sky.*—While the red rays of the sun have a greater power of penetrating a dense atmosphere, the blue rays are more readily reflected by it, but this difference is not sensible until the light has traversed large masses of air. The azure color of the sky is therefore due to the light reflected by the air, and the purer the air the more decided is this azure tint. When mountains covered with snow are illumined by a rising sun, they appear of a rosy or orange tint upon the eastern side, while on the western side they exhibit a bluish tint. The blue color of the sky is therefore due to the reflection of light, and not to a peculiar color belonging to the particles of air.

410. *Cyanometer.*—The intensity of the blue color of the sky exhibits very great variety. In order to measure it, Saussure invented an instrument which he called the cyanometer. This instrument has 27 colored surfaces, of which the first is almost white, and the last is of the deepest cobalt blue, while the intermediate surfaces present every gradation between white and blue, and the surfaces are numbered from 1 to 27. It has also a second series of colored surfaces, beginning with the deepest blue of the preceding series, and ending with a jet black, while the intermediate surfaces present every gradation between blue and black, and these surfaces are numbered from 27 to 53.

In using this instrument, the observer selects that particular tint upon the scale which corresponds nearest to the color of the sky, and the color of the sky is denoted by the number attached to that tint. Other cyanometers have been invented depending upon the properties of polarized light.

The blueness of the sky generally increases from the horizon to the zenith. When the color of the sky near the zenith is indicated by 20 on the cyanometer, it will generally be about 4 near the horizon.

The blueness of the sky is greatest after a rain, when the air is most pure, and it diminishes with an increase of the particles of condensed vapor suspended in the air. Hence a pale sky is a sign of rain.

The blueness of the sky decreases as we recede from the equator. At Cumana, in latitude 10°, the average blueness of the sky is 24, while in Europe it is only 14. On a clear, bright day, the average blueness of the sky at New Haven corresponds to about 18 on the cyanometer.

The blueness of the sky increases with the altitude, and at an elevation of 16,000 feet the heavens become almost black. On the top of Mount Blanc, Saussure found the color of the sky 39, while at the foot of the mountain the color of the sky near the zenith was represented by 18.

411. *Twilight.*—If there were no atmosphere, night would commence as soon as the sun descends below the horizon, and the day would begin with equal abruptness. The astronomical limit of twilight is generally understood to be the instant when stars of the sixth magnitude begin to be visible in the zenith at evening, or disappear in the morning. In our climate the evening twilight generally terminates when the sun is 17° or 18° below the horizon. The morning twilight commences at a somewhat less depression, since the vapor of the atmosphere condensed during the night does not rise to so great a height in the morning as at evening. These limits, however, are variable, the duration of twilight depending upon the state of the atmosphere. When the sky is of a pale color, indicating the presence of an unusual amount of condensed vapor, twilight is of longer duration. This happens habitually in the polar regions. On the contrary, within the tropics, where the air is pure and dry, twilight sometimes lasts only fifteen minutes.

412. *Twilight Curve.*—A little before sunset the western sky grows yellow, and in the east we observe a purple tint arising from the reflection of the sun's rays, which have traversed the atmosphere horizontally, and which communicate their color to whatever they illumine. After the sun has set, we perceive near the eastern horizon a dark blue segment, above which we notice the purple tint already mentioned. As the sun declines this segment rises higher; it subsequently reaches the zenith, and finally the western horizon, when the twilight entirely ceases. The outline of this segment sometimes appears very sharply defined, and is called the twilight curve. This segment is a part of the conical shadow of the earth, which intercepts the sun's rays from a portion of the atmosphere, and this portion reflects only that diffuse light which comes from other parts of the sky.

413. *Colors of the Morning Twilight.*—When the sun is still 12° below the eastern horizon, the horizon generally appears bordered with a red or orange band, above which the twilight curve rises 7°. The orange zone gradually extends, becomes bordered with yellow, and afterward with green, while the twilight curve ascends toward the zenith. When the sun is only 2° below the horizon, the eastern horizon becomes yellow, the green zone becomes more decided, and extends from 3° to 18°, the twilight curve extends to within 3° of the western horizon, and is bordered with a purple zone about 12° in breadth. As the sun rises, the western horizon appears bordered with a rosy band, surmounted by yellow. The red in the east disappears, and is succeeded by yellow, surmounted by green, which continues after the yellow has disappeared, when the sun is 2° or 4° above the horizon.

The red and yellow zones are ascribed to the absorption produced by the different thicknesses of air traversed by the rays of light. The green color results from a combination of the yellow rays with the blue rays of diffuse light reflected from the particles of the atmosphere, green being produced by a mixture of yellow and blue.

O

414. *Height of the Atmosphere deduced from Twilight.*—Attempts have been made to compute the height of the atmosphere from the position of the twilight curve at a given instant after sunset; but the results thus obtained are not uniform, being greatest when the sun is lowest below the horizon. From such computations it has been inferred that the height of the atmosphere can not exceed 36 miles; but this can only be regarded as the height of that portion of the atmosphere which has a density sufficient to reflect an appreciable amount of light. Other phenomena indicate that an extremely rare atmosphere extends to a much greater height.   .

415. *Prognostics derived from Twilight.*—Since the colors and duration of twilight, especially at evening, depend upon the amount of condensed vapor which the atmosphere contains, these appearances should afford some indication of the weather which may be expected to succeed. The following are some of the rules. which are relied upon by seamen. When, after sunset, the western sky is of a whitish yellow, and this tint extends to a great height, it is probable that it will rain during the night or the next day. Gaudy or unusual hues, with hard, definitely outlined clouds, foretell rain and probably wind. If the sun, before setting, appears diffuse and of a brilliant white, it foretells a storm. If it sets in a sky slightly purple, the atmosphere near the zenith being of a bright blue, we may rely upon fine weather.

A red sky in the morning presages bad weather, or much wind if not rain; but if the sky presents simply a rosy or grayish tint, we may expect fair weather.

SECTION III.

THE RAINBOW.

416. The rainbow consists of a series of circular bands colored with the tints of the solar spectrum from red to violet, and is situated in that part of the sky which is opposite to the sun. It is caused by the refraction and reflection of the sun's light from drops of rain whose form is sensibly spherical.

It is proved in Natural Philosophy (Olmsted, p. 419) that, if
i represents the angle of incidence of a ray of light,
r "   "   "   refraction   "   "

D represents the angle of deviation of a ray of light,

n " the index of refraction for water;

then, for the maximum deviation after one reflection, we have

$$\cos. i = \sqrt{\frac{n^2-1}{3}}\ ; \ \sin. i = n \sin. r\ ; \ D = 4r - 2i.$$

If we assume the index of refraction for the red rays to be 1.3309, and for the violet rays 1.3442, we shall find

for the red   rays, $i = 59°\ 32'$; $D = 42°\ 24'$;

"   violet rays, $i = 58°\ 46'$; $D = 40°\ 28'$.

For the minimum deviation after two reflections we have

$$\cos. i = \sqrt{\frac{n^2-1}{8}}\ ; \ \sin. i = n \sin. r\ ; \ D = \pi + 2i - 6r.$$

Whence, by computation, we find

for the red   rays, $i = 71°\ 55'$; $D = 50°\ 20'$;

"   violet rays, $i = 71°\ 29'$; $D = 53°\ 46'$.

The exterior radius of the primary bow should therefore be 42° 24', increased by half the diameter of the sun; and its breadth should be 1° 56', increased by the apparent diameter of the sun, which is about 30', making 2° 26'. The mean of numerous careful measurements gives 41° 33' for the radius of the middle part of the primary bow.

The interior radius of the secondary bow should be 50° 20', diminished by half the diameter of the sun, and its breadth should be 3° 26'+30', or 3° 56'.

417. *Necessary Conditions of Visibility.* — If the altitude of the sun be greater than the radius of the bow, then no rainbow can be seen. For this reason, during more than six months of the year at New Haven, the primary bow can never be seen at noon, and near the summer solstice the primary bow can not be seen for more than six hours near the middle of the day.

If the observer be sufficiently elevated above the earth, as in a balloon, he may see the rainbow as a complete circle, but on the surface of the earth we only see a semicircle when the sun is in the horizon.

Lunar rainbows are occasionally seen, but the colors are faint, and generally only a white or yellowish arc is distinguishable.

418. *Supernumerary Bows.* — The Newtonian theory of the rainbow is incomplete, inasmuch as it only considers those rays which experience the maximum or minimum deviation, and entirely neglects those rays which pass a little beyond these limits. The effect of these other rays is to extend the breadth of the primary bow upon the inside, and also to produce secondary bands which the Newtonian theory does not explain. When the rainbow is brilliant, we often perceive faint bands alternately red and green within the violet of the primary bow, or perhaps superposed upon the violet, which then assumes a purplish tint. Near the violet bow we frequently see an arch of rose-red, succeeded by one of yellowish-green; then perhaps a second arch of rose-red and a second of yellowish-green. Two supernumerary bows are not very uncommon; three have repeatedly been seen, and occasionally even four.

These supernumerary bows are due to the interference of rays which traverse a drop in a direction differing but little from that of maximum deviation. To every angle of deviation a little less than the maximum, there correspond two rays, one whose angle of incidence is a little greater, and the other whose angle is a little less than that which gives the maximum deviation. These rays, having pursued routes slightly unequal, interfere and produce alternations of light and darkness, or alternately bright and dark bands. The bands resulting from these interferences for each of the colors of the spectrum, being superposed upon the sky, produce bands analogous to the colored rings of thin plates.

419. *Theory explained from a Diagram.*—A ray of light, SA, once reflected from the inner surface of a drop of rain at B, ex-

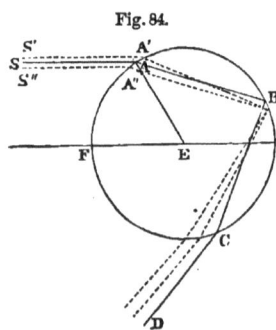
Fig. 84.

periences its greatest deviation, viz., 41°, when the angle of incidence, FEA, is 59°. Suppose a ray of light, S′A′, falls upon the drop at an angle greater than 59°, it will experience a deviation less than 41°. So also a ray, S″ A″, which falls upon the drop at an angle less than 59°, will experience a deviation less than 41°. That is, there are always two rays which experience an equal deviation (for example, 40°), and therefore emerge parallel, one of them making with the drop an angle greater than 59°, and the other a less angle. The paths of these two rays within the drop are slightly unequal, and there are two rays the difference of whose paths within the drop is equal to half the breadth of a wave of light. These waves, being in opposite phases, will interfere with each other, and produce darkness. There are two other rays the difference of whose paths within the drop is equal to the breadth of a wave of light. These waves, being in the same phase, will conspire to produce a double illumination. There are two other rays the difference of whose paths is equal to one and a half undulations, and which consequently interfere with each other.

Thus we have rays the difference of whose paths is equal to 1, 2, 3, 4, etc., undulations, and which therefore conspire; and there are other rays the difference of whose paths is equal to $\frac{1}{2}, 1\frac{1}{2}, 2\frac{1}{2}, 3\frac{1}{2}$, etc., undulations, and which therefore interfere.

420. *Consequence of these Interferences.*—If, then, the sun furnished red light only, we should see opposite to the sun, when drops of rain are falling, circular arcs, alternately red and black. If the sun's light were entirely violet, we should see circular arcs alternately violet and black, but the diameter of the violet arcs would be less than that of the red arcs. The other colors of the spectrum would produce arcs of intermediate dimensions. Now, since the sun's light contains all the colors of the spectrum, all these colored arcs are in fact formed simultaneously and superposed, and, being of unequal diameters, the colors are partially blended. But near the usual primary bow two or three of these narrow bands of prismatic colors are often sufficiently distinct to be visible. In consequence of this reflection of light from the drops of rain, it results that the sky within the primary bow is brighter than that without it.

421. *Size of the Drops of Rain.*—The smaller the drops of rain, the broader will be these colored bands. In order that the supernumerary bows may be formed beyond the first violet bow, the drops must be extremely minute. It is found by computation that if the drops be $\frac{1}{77}$ inch in diameter, a second red band will be formed 2° from the outer red of the primary bow, and it is near this point that the first supernumerary bow is usually seen.

If we consider the interval between the first and second maximum unity, the breadths of the succeeding intervals for the same color will be expressed by the numbers,

second interval, 0.587 ;   fourth interval, 0.440 ;
third interval, 0.493 ;   fifth interval, 0.404.

Supernumerary bows are sometimes seen on the outside of the secondary rainbow, and they are to be explained in a similar manner.

422. Fog-bow explained.—If the drops be less than $\frac{1}{77}$ inch in diameter, the primary bow will be wider than two degrees, the breadth of the bow depending simply upon the size of the drops. But as the breadth of the bow increases, the colors are spread over a greater surface, and consequently they are less vivid and distinct. When the diameter of the drops is $\frac{1}{130}$th inch, which is the average diameter of particles of fog, the bow becomes a very faint arch 4° or 5° in breadth, with only a slightly rosy tint upon the outside. Such is the bow actually observed when the sun shines upon a dense fog.

The undulatory theory of light, therefore, explains not only the supernumerary bows, but the variable breadth of the primary bow.

SECTION IV.

CORONÆ.

423. The sun and moon, when partially covered by light, fleecy clouds, are often seen encircled by one or more colored rings, which are called *coronæ*. This phenomenon is most frequently noticed about the moon, since we are too much dazzled by the light of the sun to distinguish faint colors surrounding his disc. In order to examine coronæ about the sun, it is best to view them by reflection from a blackened mirror, by which means the brilliancy of the sun's light is very much reduced.

424. Order of the Colors.—When a corona is complete, we may observe several concentric colored circles. The one next to the sun is blue, the second is nearly white, and the third is red. These form the first series of rings. In the second series the order of the colors is purple, blue, green, pale yellow, and red. In the third series the colors are pale blue and pale red. These rings are partially represented in Fig 88.

The diameter of these rings is not always the same. The diameter of the first red ring varies from 3° to 6°, and that of the second red ring from 5° to 10°.

425. *Cause of Coronæ.*—Coronæ are produced by the diffraction of the rays of light in their passage through the small intervals between the particles of condensed vapor in a cloud. If we look at the moon through a very small aperture (as a pinhole in a plate of sheet-lead) we shall see the hole surrounded by colored rings, whose tints are the same as those observed in coronæ. The light of the moon, passing through the small interstices between the particles of a cloud, is diffracted in a similar manner. The particles of a cloud must not be too numerous, otherwise no rays can pass between them; and the smaller the intervals between the particles, the greater will be the diameter of the rings.

426. *Coronæ produced artificially.*—If we sprinkle upon a pane of glass a little lycopodium, or any very fine dust of nearly uniform fineness, and look at the moon through this glass, we shall see it surrounded by rings of the prismatic colors, precisely like those formed by a cloud.

If, on a cold winter evening, we breathe upon a pane of glass, the breath will condense in small globules and freeze; and if we look at the moon, or even at a street lamp, through this glass, we shall see a similar system of colored rings, having violet on the inside.

427. *Glow surrounding the Shadow of an Observer.*—When the sun is near the horizon, and the shadow of the observer falls on grass covered with dew, one may often observe a vivid glow surrounding the shadow of his head. If the shadow falls upon a cloud or a fog, the head will appear surrounded by a luminous glory, exhibiting the prismatic colors. The order of the colors is the same as in coronæ, and sometimes four and even five series of rings have been observed.

The light of the sun is reflected to the eye most powerfully by the particles of fog near the head; for the light reflected both from the anterior and posterior face of such particles will reach the eye. This explains the glow of light surrounding the shadow of the observer. The *color* is produced by the diffraction of the light thus reflected, precisely as in the case of a corona.

SECTION V.

HALOS AND PARHELIA.

428. Halos are circles of prismatic colors formed around the sun or moon. They are of larger size than coronæ, and present a greater variety of appearances. The following is an enumeration of those which are most frequently seen.

Halo of 22° Radius. — When the sky is hazy, and presents a dull, milky appearance, we frequently notice around the sun or moon a colored circle, *h*, Fig. 85, having a radius of 22°, the sun

Fig. 85.

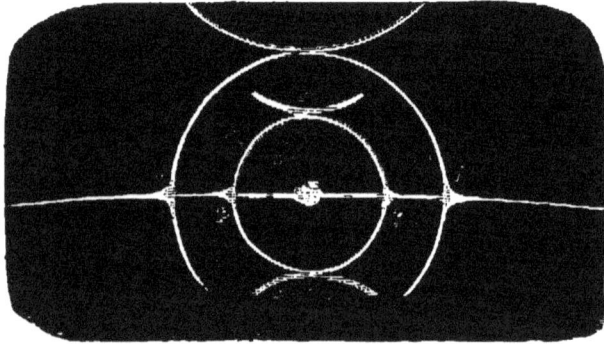

occupying the centre of the circle. The inner edge of the circle is colored red, and is tolerably well defined; the outer edge is of a pale blue, and is not sharply defined. Such a circle is never seen when the sky is perfectly clear. The sky within the halo is much darker than it is for a distance of several degrees without the halo.

The light of this halo is always polarized in the direction of a tangent to the circumference, which proves that its light has suffered refraction and not reflection.

429. *Theory of this Halo.*—This halo is formed by the refraction of the light of the sun or moon through crystals of ice floating in the atmosphere. Snow consists of crystals of ice. The simplest form of an ice-crystal is a right prism, whose section is a regular hexagon, and terminated by two bases perpendicular to the edges of the prism. The alternate faces of such a prism are inclined to

Fig. 86.

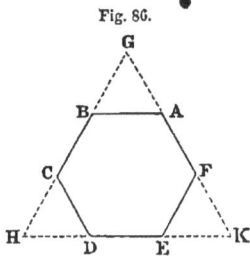

each other at angles of 60°, so that we may consider the hexagonal prism ABC DEF as a triangular one, GHK, with angles of 60°.

When a ray of light passes through a prism, it is deviated toward the base of the prism; and there is a certain position of the prism in which the deviation is the least possible. This deviation for a prism of ice may be computed in the following manner:

Fig. 87.

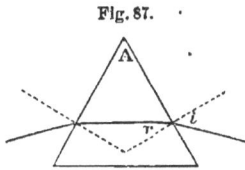

Let i represent the angle of incidence of a ray of light; r the angle of refraction; m the index of refraction; and A the refracting angle of the prism. Then

$$\sin. i = m \sin. r.$$

But when the deviation is a minimum, $r=30°$; and for red light the value of m is 1.307. Hence $i=40°$ 48½'. And the deviation of a ray of light is

$$2i - A, \text{ which equals } 21° 37'.$$

The minimum deviation for the violet rays, for which the value of m is 1.317, is found, in like manner, to be 22° 22'.

430. *How a Circle of Light is formed.*—If we conceive a beam of light to be admitted through a small aperture into a dark room, and to fall upon a large number of ice prisms having angles of 60°, and occupying every possible position, all the incident rays will be deviated from their first direction, but in no case will the deviation be less than about 22°. A large number of spectra will be cast upon the opposite wall, but opposite to the aperture through which the light is admitted there will be a circle of 22° radius upon which no spectrum can fall, and the red end of each spectrum will be turned toward the centre of the circle. If the number of the spectra be sufficiently great, they will together form a circle of 22° radius, bordered with red upon the inside; but beyond the red the different colors of the spectrum will be so superposed as to produce a light nearly white.

Whenever halos are formed about the sun, the air is filled with fine prismatic crystals of ice, and these crystals occupy every possible position with respect to the sun's light. The halo of 22° radius is formed by the light of the sun shining through these crystals of ice. If the sun's light furnished only red rays, we should have an illuminated surface with a circular opening of $21\frac{1}{2}°$ radius, of which the inner edge would be quite light. If the sun furnished only violet rays, we should have a similar violet surface, with a circular opening of $22\frac{1}{2}°$ radius, and the intermediate colors would furnish circles of intermediate dimensions. Now, since the sun's rays contain all the colors of the spectrum, these different circles are formed simultaneously and superposed. The red projecting on the inside is unmingled with any other color, and is therefore pure; all the other colors are more or less mingled, but in unequal proportions, so that the outer portion of the halo is nearly white.

Such a halo may be formed in midsummer, because at a moderate elevation above the earth's surface the condensed vapor of the air is frozen even in the hottest weather. The circle within the halo is much darker than the space without it, because from no part of this circle can a ray of the sun refracted by ice prisms reach the eye of the observer.

The mean of eighty-three measurements of the radius of the red circle of this halo is 21° 36′, which is almost identical with the radius computed from theory.

431. *Halo of 46° Radius.*—Sometimes we notice around the sun a second colored circle, H, Fig. 85, having a radius of 46°. The inner edge of this circle is also red, and tolerably well defined, while the outer edge is of a pale blue color, and is poorly defined. This halo is formed by the refraction of the sun's rays through ice prisms having an angle of 90°, this being the angle which each side of the hexagonal prism forms with its base. The minimum deviation of a ray of red light through a prism of ice having such a refracting angle is found by computation to be 45° 6′, and for a ray of blue light 46° 50′. The average of the best observations give 45° 46′ as the radius of the brightest part of this halo, a coincidence as exact as can be expected in observations of this nature.

432. *Halos produced artificially.*—The production of halos may be experimentally illustrated by crystallizing some salt like alum upon a glass plate, and then looking through the plate at the sun or a candle. A few drops of a saturated solution of alum spread over a plate of glass will soon cover it with a layer of minute crystals. If we place the eye close behind the smooth side of the glass, and look at a candle, we shall see the candle surrounded by three halos of different dimensions. Each crystal of alum is a regular octaedron, with the six angles truncated, forming the outline of a cube. It has therefore faces inclined to each other at angles of 70°, 90°, and 110°, and these angles occupy every possible position with respect to the glass plate. The smallest halo is formed by the refraction of the rays of light through a pair of faces inclined to each other at an angle of 70°; the second halo is formed by a pair of faces inclined to each other 90°; and the third halo by faces inclined at an angle of 110°.

433. *Halo of* 90° *Radius.*—A third halo of about 90° radius, H′, Fig. 88, is occasionally seen surrounding the sun. Unlike the other two halos, this halo shows scarcely any traces of the prismatic colors. Only three observations of this halo are on record,

Fig. 88.

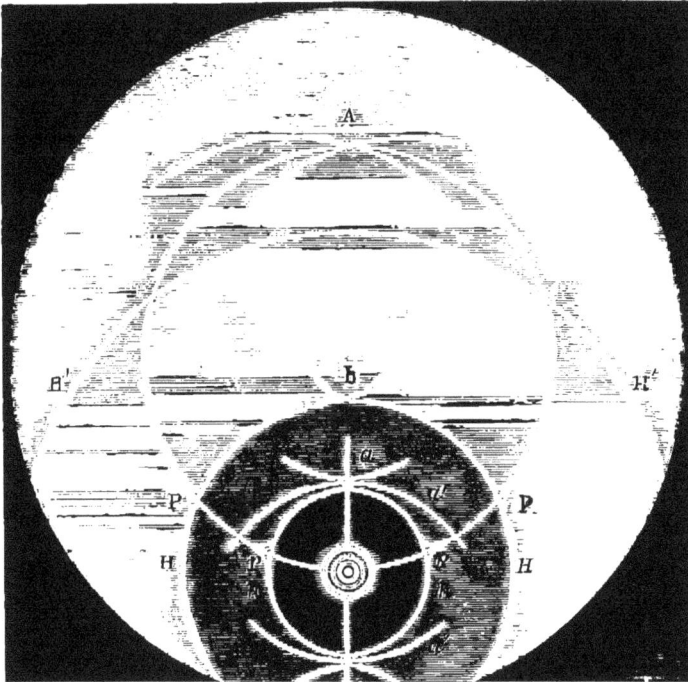

and its exact dimensions have not been well determined. In two of the observations the radius was estimated at 90°, and in the third it was estimated from 85° to 90°.

This halo has been ascribed to rays which, after entering one of the sides, AB, of a triangular ice prism, meet the face, BC, at such an angle that they are totally reflected, and emerge through the face AC. The angle of total reflection, r, is determined by the equation

Fig. 89.

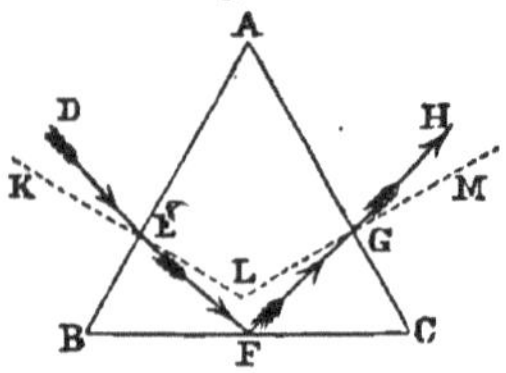

$$\sin. r = \frac{1}{m}.$$

For violet rays in ice, $m = 1.317$; whence $r = 49° 24'$, or BFE $= 40° 36'$. Hence FEL $= 10° 36'$.

Also KED $= m \sin. $ FEL $= 14° 1'$.

The inclination of DE to GH is equal to $120° - 2.$ KED $= 91° 58'$.

Such a reflection from an indefinite number of ice prisms would therefore furnish an illumined surface with a circular opening of about 92° radius, and having a tinge of violet on the side next to the sun. The radius above computed is somewhat greater than that indicated by the observations, and there are other objections to this explanation, so that this hypothesis is quite doubtful; but no satisfactory explanation of this halo has hitherto been proposed, and the observations are not sufficiently precise to enable us to choose between conflicting hypotheses.

434. *Parhelic Circle.*—When a halo is formed around the sun we often notice a white circle passing through the sun and parallel to the horizon. See Fig. 88. This is called a *parhelic* circle, and is produced by the reflection of the sun's light from ice prisms or snow crystals whose surfaces have a vertical position. When the air is tranquil, the flakes of snow which are present in the atmosphere descend slowly to the earth, and they tend to assume that position in which they experience the least resistance from the air. For most forms of snow-flakes, this position will be when the principal faces of the crystal are perpendicular to the horizon, and the light of the sun may reach the eye reflected from such snow-flakes as are situated on a horizontal circle passing through the sun. This circle never exhibits prismatic colors like the first-mentioned halos.

435. *Parhelia.*—Near those points where halos cut the parhelic circle there is a double cause of light, and here the illumination is sometimes so great as to present the appearance of a mock sun, and is called a parhelion. Parhelia are generally red on the side which is toward the sun, and they sometimes have a prolongation in the form of a tail several degrees in length, and whose direction coincides with that of the horizontal circle.

Parhelia of 22°.—The *number* of parhelia is very variable. One is commonly seen near each of the points where the parhelic circle cuts the halo of 22° radius, *pp*, Fig. 88, but the distance of this parhelion from the sun increases with the elevation of the sun above the horizon. When the atmosphere is calm, the prisms of ice which are present in the air, and are slowly descending to the earth, will tend to assume a vertical position; and if the sun be near the horizon, the brightness of this halo will be greatest at each extremity of a horizontal diameter. As the sun rises above the horizon, the rays of light traverse these vertical prisms in a direction oblique to the axis, and the minimum deviation of a ray is increased, and the parhelion recedes from the circumference of the halo. For an elevation of 20° this deviation amounts to a degree and a quarter; at an elevation of 40°, it amounts to more than five degrees; and at an elevation of about 50°, this parhelion entirely disappears on account of the oblique angle at which the rays meet the ice prisms.

Parhelia of 46°.—A parhelion is sometimes seen at each of the points PP, Fig. 88, where the parhelic circle cuts the halo of 46° radius. These parhelia have never been seen to depart much from the circumference of the halo; but since the breadth of the halo is 1½°, and that of the parhelion is still greater, it is not certain that the coincidence is exact. These parhelia can not be ascribed to ice prisms with angles of 90°, the edges of these angles being vertical, for such a position of the base of an hexagonal prism would be unstable. Moreover, upon such an hypothesis, as the sun rises above the horizon, the parhelion ought to recede rapidly from the halo of 46°, which is contrary to observation.

These parhelia are probably produced by rays which have experienced the minimum deviation in the same direction in two vertical hexagonal prisms, in which case the total deviation of the rays would be double of that produced by a single prism. Upon this hypothesis the parhelia should not exactly coincide with the halo of 46°, but for elevations not exceeding 30° the difference might easily escape observation. The observations are not sufficiently precise to decide whether this explanation is admissible or not.

Parhelia of 120°.—Two other parhelia are sometimes seen on the parhelic circle, about 120° distant from the sun. These may

Fig. 90.

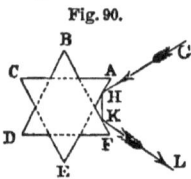

be caused by two reflections of the rays of the sun from the vertical faces of snow crystals, whose form is such as is represented by Fig. 90. The ray GH, after two reflections at H and K, takes the direction KL, experiencing a total deviation of 120°. The image formed by this reflection is white, and its size about equal to that of the sun's disc.

Parhelia have also been observed at distances of 50° and 98° from the sun, which may result from the reflection of the sun's rays from the faces of snow crystals of more complicated forms.

Sometimes a parhelion is seen on the parhelic circle at A, Fig. 88, directly opposite to the sun. This is more properly called an *anthelion.*

Phenomena similar to parhelia are produced by the light of the moon, in which case these bright spots are called *paraselenœ.*

436. *Contact Arches.* — Arcs of colored circles with variable curvatures are sometimes seen touching the halos of 22° and 46° at their highest and lowest points, *a*, *b*, Fig. 88. These are due to the refraction of the sun's light through ice prisms, some of them having their axes perpendicular to the sun's rays, and others inclined at various angles, but all in a horizontal position. The sun's light, refracted by such prisms as have their axes not only horizontal, but perpendicular to the solar rays, will produce a bright image directly over or under the sun. But the sun's light, passing through prisms whose axes are inclined to the solar rays, will experience a greater deviation, and also a deflection from a vertical plane. Thus, if we look at a long straight bar through a prism whose axis is parallel to the bar, the straight bar appears curved, the deviation being greatest in the case of those rays which are oblique to the axis of the prism.

437. *Variable form of Contact Arches.*—The form of these contact arches depends upon the height of the sun above the horizon. When the sun is near the horizon, we sometimes see two brushes of light, like horns, rising from that point in the halo of 22° which is directly over the sun. As the sun rises higher, these two horns diverge from each other, and when the sun has an altitude of 12°, they approach in form to an arc of a circle, with its convexity toward the sun. When the sun reaches an altitude of 30°, these arcs become concave toward the sun; they bend downward, and partially envelop the halo.

When the sun has an altitude of 25°, a contact arch is sometimes seen at the point of the halo directly beneath the sun. At first it appears like an arc of a circle, with its convexity turned toward the sun. As the sun rises higher, the curvature of this arc diminishes, and at an altitude of 32° the arc becomes concave toward the sun. At the height of 45° the curvature of the lower contact arch is nearly the same as that of the upper arch, and

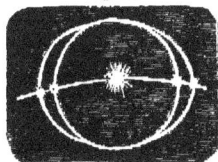

Fig. 91.

both together form an elliptical figure, surrounding the halo of 22°, as shown in Fig. 91. When this ellipse is greatest, the length of its horizontal axis is about 64°. As the sun rises still higher, the major axis of this ellipse contracts, and when the sun's altitude is 60°, the horizontal axis of the ellipse is reduced to 50°. At the altitude of 70°, the ellipse differs so little from the halo itself as to be scarcely distinguishable from it.

All these arcs are due to the sun's light, refracted by ice prisms having their axes horizontal, as may be verified experimentally by passing the sun's light through a triangular water prism held in the proper position with respect to the sun's light.

438. *Arcs touching the Halo of 46°.*—When the sun has an altitude of 12°, a brilliant arch in the form of an inverted rainbow is sometimes seen to touch the halo of 46° at its highest point, b, Fig. 88. As the sun rises higher in the heavens this arc becomes more curved, and it disappears when the sun attains an altitude of 31°.

When the sun has an altitude of 60°, a colored arch is sometimes seen touching the halo of 46° at its lowest point; but its light is faint, and it is concave toward the sun, so that this arc is easily confounded with the halo itself. As the sun rises higher in the heavens, this arch approaches still nearer to coincidence with the halo, and it disappears entirely when the altitude of the sun is 78°.

These arcs are formed by the refraction of the sun's light through ice prisms having angles of 90°, the edges of these angles being situated in a horizontal plane; and the angles will have this position when the axis of the hexagonal prism is vertical. A ray of the sun can not pass through so large a refracting angle except when the sun has a particular altitude above the horizon. It is for this reason that the upper contact arch is never seen except when the sun's altitude is between 12° and 31°; and the lower contact arch is never seen except when the sun's altitude is the complement of the preceding, viz., from 59° to 78°.

439. *Intersecting Arcs opposite to the Sun.*—Sometimes we notice two arcs of circles nearly white, A, Fig. 88, intersecting the parhelic circle at a point directly opposite to the sun, and inclined to this circle at angles of about 60°.

· They are probably due to reflection from surfaces oblique to the horizon; perhaps from the slender spiculæ of snow-flakes having

Fig. 92.   Fig. 93.

the form and position shown in Fig. 92, or from hexagonal snow-plates whose bases are covered with striæ arising from the superposition of other hexagonal plates in the manner shown in Fig. 93. Flakes of snow having such a figure have been repeatedly observed.

440. *Vertical Columns passing through the Sun.* — Sometimes, near sunset, we notice a luminous column, perpendicular to the horizon, rising from the sun to a height of 10° or 15°, and occasionally still higher. This column is due to the reflection of the sun's light from the under faces of ice crystals which are nearly parallel to the horizon. Sometimes a little before sunset a similar column of light is seen to shoot down from the sun toward the horizon. This is formed in a similar manner by rays of the sun reflected from the upper faces of crystals in a nearly horizontal position. Sometimes columns are seen simultaneously both above and below the sun; and if the halo of 22° is seen at

Fig. 94.

the same time, this column, together with the parhelic circle, presents the appearance of a rectangular cross within the halo, Fig. 94. These luminous columns are probably formed only when the air is very tranquil, and the reflecting surfaces may be the rectangular terminations of spiculæ of ice which are slowly falling to the earth, with their axes nearly in a vertical position.

When we remember the immense variety in the forms of snow-flakes, a few of which are represented in Fig. 52, we should anticipate a very great variety in the figures which might be produced from the refraction or reflection by them of the sun's light. In addition to the figures which have been described in this section, many others have occasionally been seen, but the descriptions which have been furnished of them are not, in general, sufficiently precise to enable us to decide respecting their proper explanation.

CHAPTER IX.·
SHOOTING-STARS, DETONATING METEORS, AND AEROLITES.
SECTION I.
SHOOTING-STARS.

441. *Shooting-stars described.*—The term shooting-star, or falling-star, is employed to designate that luminous stream which at night is frequently seen to shoot rapidly across the sky, and presently vanishes, appearing as a star which is shot away from its place in the firmament to a distant region of the heavens. Shooting-stars may be seen on every clear night, and at times follow each other so rapidly that it is quite impossible to count them.

442. *Number seen at different Hours.*—Shooting-stars are not seen with equal frequency at all hours of the night. They generally increase in frequency from the evening twilight throughout the night until the morning twilight; and when the light of day does not interfere, they are generally most numerous about 6 A.M. The following table shows the average number of shoot-

P

ing-stars which may be seen by a single observer at each hour of a clear night, in the absence of the moon:

·From	6 to	7 P.M.,	3.8		From 12 to	1 A.M.,	7.2
"	7 "	8 "	3.8		" 1 "	2 "	7.8
"	8 "	9 "	4.0		" 2 "	3 "	8.7
"	9 "	10 "	4.7		" 3 "	4 "	10.3
"	10 "	11 "	5.3		" 4 "	5 "	11.2
"	11 "	midnight,	6.0		" 5 "	6 "	11.2

Observations show that the whole number of shooting-stars visible at one place must be at least six times the number which can be seen by one observer. Hence the average number of meteors that traverse the atmosphere, and that are large enough to be visible to the naked eye, if the sun, moon, and clouds would permit, is 42 in an hour, or 1000 daily.

443. *Number seen in the different Months.* — Shooting-stars are not seen with equal frequency at all seasons of the year. The following table shows the average hourly number which may be seen by a single observer near midnight, during each month, on clear nights, in the absence of the moon:

January .	. 5.1	May . . .	4.0	September .	7.4
February .	5.0	June . . .	4.9	October . .	7.7
March . .	4.8	July . . .	10.0	November .	7.4
April . .	4.6	August . .	12.8	December .	5.4

We thus see that many more shooting-stars appear from July to December than during the other six months of the year; and they are ordinarily most abundant in the month of August.

444. *Altitude of Shooting-stars.* — If two observers, at a suitable distance from each other, note the apparent altitude and azimuth of a shooting-star at the commencement of its flight, and do the same also for its termination, we have the data for computing the absolute height of beginning and end above the surface of the earth. The earliest observations of this kind were made in 1798 by Benzenberg and Brandes in Germany, and since that time similar observations have been made in many parts of Europe, and also in the United States. It is found that when the base-line employed is only three or four miles in length, a shooting-star is seen in nearly the same direction at both stations, showing that its altitude is much greater than the length of that base.   When the base-line is 30 or 40 miles, the average change of position of the star is about 15°.   The base-line should not be less than 40 or 50 miles in length, and one of 75 or 100 miles would not be too great.   Observers at distances of over 150 miles see for the most part different shooting-stars.

The heights of over 500 meteor paths have been computed, and it is thus found that shooting-stars begin to be visible at elevations of from 40 to 120 miles, and perhaps sometimes 150 miles, or an average height of 74 English statute miles.   They disappear at elevations of from 30 to 80 miles, and perhaps sometimes 100 miles or more, giving an average height at disappearance of 52 English statute miles.

445. *Length of Path and Velocity.* — The length of the visible path of shooting-stars varies from 10 to 100 miles, though sometimes they are even 300 and 400 miles long; the average length being 28 miles.   The time of describing the visible path varies from less than one second to five seconds, and in some rare cases amounts to ten seconds; but their average duration is less than one second.   The average duration of meteors whose brightness exceeds that of stars of the first magnitude is estimated at one and a half seconds.

Their velocity relative to the earth's surface varies from 10 to 45 miles per second, and the average velocity of the brighter class of shooting-stars amounts to about 30 miles per second.

446. *Direction of their Motions.* — Shooting-stars are seen to move in all directions through the heavens.   Their apparent paths are, however, generally inclined downward, though sometimes they move upward, and after midnight they come in the greatest numbers from that quarter of the heavens toward which the earth is moving in its annual course around the sun.

447. *Magnitude of Shooting-stars.*—The magnitude of shooting-stars is very variable.   Some of them have been computed to have a diameter of 100 or 200 feet, and others 1000 up to 5000 or 6000 feet.   We must, however, regard this as the diameter of the blaze of light which surrounds the meteor, while the meteor itself, before it takes fire, may have a diameter of only a few feet, or perhaps only a fraction of an inch. The apparent size of meteors is greatly magnified by irradiation.

448. *Visible Train.*—Occasionally shooting-stars appear in great splendor, flashing with a brightness nearly equal to that of the full moon, and leaving behind them a train of dazzling light, which lasts for several seconds, and even for whole minutes. Their color is usually white, with a reddish tinge; but occasionally they exhibit a green light, and sometimes a mixture of green and blue, or purple. Even quite faint shooting-stars sometimes leave trains.

The path of shooting-stars is frequently curved—sometimes the path consists of two portions inclined to each other at a considerable angle—and at the end the meteor sometimes bursts like a rocket into numerous fragments. In such cases the place of explosion is usually indicated by a smoky cloud, which sometimes continues visible for ten minutes.

449. *Are Shooting-stars accompanied by any Sound?*—Observers frequently imagine that they hear a whizzing noise accompanying the passage of a brilliant meteor. It is easily proved that such impressions are an illusion. When we compute the path of the meteor, from which the sound was supposed to proceed, we always find that it was quite distant from the observer, 20, or 50, and perhaps 100 miles. Now sound moves with a velocity of 1120 feet per second, or 50 miles in about four minutes. If, then, any noise was caused by the motion of the meteor, the sound could not possibly be heard until considerable time after the meteor disappeared, viz., 2, 5, or even 10 minutes, according to its distance.

450. *Cause of the Light of Shooting-stars.*—This light is probably due to the high temperature resulting from the resistance of the atmosphere to the rapid motion of the meteor. Since, at the ordinary elevation of shooting-stars, the air is exceedingly rare, it might be supposed that the resistance would not develop sufficient heat to give meteors their brilliant appearance. The researches of philosophers have enabled us to compute the quantity of heat that may be developed by the stoppage of a meteor in the atmosphere. A portion of the living force of the body is expended in setting the air in motion, and a portion in heating the meteor and the air. This living force and the consequent heat that may be developed is proportioned to the mass of the body and to the square of its velocity. The arresting the motion of a meteor whose velocity is thirty miles per second, and whose specific heat is 0.12, would, if the whole living force were changed into heat, be sufficient to raise the temperature of the meteoric body more than four million degrees of Fahrenheit's scale. If even the larger part of this force was expended in giving motion to the air, there would remain enough to furnish a brilliant light, and to melt or disintegrate the meteor.

451. *Daily number of Shooting-stars for the whole Globe.*—The mean distance of shooting-stars from the observer is found to be about 105 miles. The average height above the earth of the middle points of their paths is 63 miles. Hence the mean horizontal distance of the paths may be regarded as about 90 miles. It is reasonable to suppose that the number of shooting-stars actually falling within a circle of 90 miles radius is at least equal to the number seen at one place. In fact, careful computations show that it is about one fourth greater. The area of this circle is 25,447 miles, while the entire surface of the globe is 196,662,000 square miles. The ratio of these numbers is 7728, whence we may safely conclude that the number of shooting-stars over the whole earth is more than eight thousand times the number visible at one place.

The average daily number of shooting-stars visible to the naked eye at one place has been estimated at 1000, Art. 442. Hence the average number of meteors that traverse the atmosphere daily, and that are large enough to be visible to the naked eye, if the sun, moon, and clouds would permit, must be more than 1000 × 8000, or more than *eight millions.*

452. *Number of telescopic Shooting-stars.*—The observations of Pape and Winnecke indicate that the number of meteors visible through the comet-seeker employed by the latter is about 40 times the number visible to the naked eye. A further increase of optical power would doubtless reveal a still larger number of these small bodies. Hence we must conclude that the source from which these meteors come is of immense extent, otherwise it would long since have been exhausted.

The mass of these bodies is, however, so small, and their distance from each other so great, that they exert no appreciable influence upon the motion of the planets. It is computed that the average distance from each other of shooting-stars, such as under favorable circumstances would be visible to the naked eye, is about three hundred miles.

453. *Meteoric Orbits.*—Having determined the velocity and direction of a meteor's path with reference to the earth, and know: ing, also, the direction and velocity of the earth's motion about the sun, we can compute the direction and velocity of the motion with reference to the sun. This computation has been made for several different meteors, and has shown that these bodies, before they approached the earth, were revolving about the sun in ellipses of considerable eccentricity. In some instances the velocity has been found to be so great as to indicate that the path differed little from a parabola.

It is thus demonstrated that ordinary shooting-stars are small meteoric bodies, moving through space in paths similar to the comets, and it is probable that they do not differ materially from the comets except in their dimensions, and perhaps, also, in their density.

454. *Periodic Meteors of November.*—We have seen, Art. 443, that the average number of shooting-stars for the different months of the year is quite unequal, and occasionally the display of meteors is very extraordinary. The most remarkable exhibitions of this kind during the last two centuries have occurred in November. On the morning of November 13, 1833, throughout most of North America, shooting-stars appeared in such numbers that it was found impossible to count them. At Boston it was estimated that the meteors fell at the rate of 575 per minute. Most of these meteors moved in paths which, if traced backward, would meet in a single point, or small area, situated near γ Leonis.

On the 13th of November, in 1832, shooting-stars appeared in very unusual numbers, and there was a moderate display on the same day of 1834, 1835, and 1836.

On the morning of November 12th, 1799, an extraordinary fall of shooting-stars was witnessed in South America by Humboldt, and it was also seen throughout a considerable part of North America. The examination of old historical records has led to the discovery of at least ten other similar appearances at about the same season, of the year. These occurred in the years 902, 931, 934, 1002, 1101, 1202, 1366, 1533, 1602, and 1698.

455. *Meteoric Shower of November 14th,* 1866.—These remarkable displays having occurred at intervals of 33 or 34 years, or some multiple of that period, led to a general expectation of a brilliant shower in 1866. At New Haven, on the night of November 13th–14th, 881 meteors were counted in five hours, which is six times the average number for November; but a far more brilliant display was witnessed in Europe. On the morning of November 14th, at Greenwich, from midnight to 1 o'clock, there were observed 2032 meteors; from 1 to 2 o'clock, 4860 meteors; and from 2 to 3, 832 meteors, the maximum occurring about a quarter past one, when the number amounted to 120 per minute. The curve line, Fig. 95, shows the number of meteors observed ·

Fig. 95.

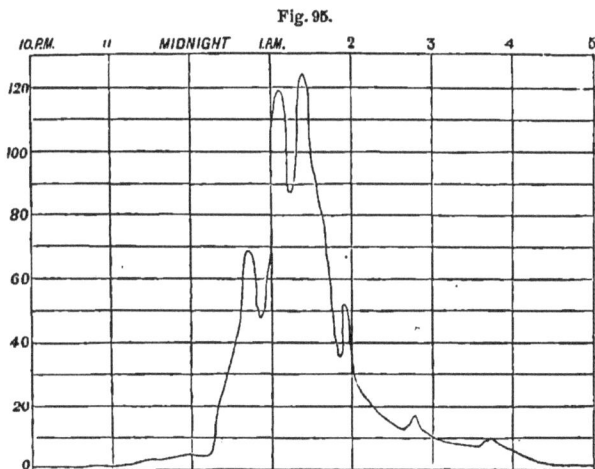

each minute from 10 P.M., November 13th, to 5 A.M., November 14th, the number visible at each instant being indicated by the numerals 0 to 120 on the left of the diagram. Nearly all of these meteors proceeded from a point in the constellation Leo; this point being in latitude 10° N., and its longitude being about 90° less than that of the sun.

A similar display was noticed throughout Europe; also in Asia as far eastward as Calcutta, and in corresponding longitudes in the southern hemisphere. Throughout all this region the maximum display occurred at about the same instant, of absolute time.

456. *Meteoric Shower of November 14th*, 1867.—An equally remarkable display of meteors occurred in the United States on the morning of November 14th, 1867. Until 3 A.M. the number of shooting-stars was not remarkable, but from that hour the number rapidly increased, and at New Haven attained its maximum about 4½ A.M., after which the number declined, and before six o'clock had ceased to be specially noticeable. Near the time of maximum the number visible to a single person was 43 per minute, making about 240 per minute for the entire heavens, and this in the presence of a full moon, which probably eclipsed two thirds of those which would otherwise have been visible. These meteors almost without exception moved in paths which, if produced backward, would intersect, not all precisely in a single point, but within a small area situated in Leo. This area was of an oval form, having a diameter of about 5° in longitude and 1° in latitude. Its centre was in longitude 143°, and latitude 10° 10' N., and most of the meteors appeared to diverge pretty accurately from this centre. Many of them left trains which were distinctly visible for several seconds, notwithstanding the light of the moon.

457. *Procession of the Node along the Ecliptic.*—The day of the year upon which the great displays of the November meteors occur becomes gradually later and later. In 1866 and 1867 the great display was November 14th; in 1832 and 1833 it was November 13th; in 1799 it was November 12th; in 1698 it was November 9th; and the earliest recorded corresponding displays occurred in October. If we suppose that these meteors, before they encounter the earth, form a ring, or a portion of a ring, about the sun, then we must conclude that the node of this ring has a direct motion along the ecliptic amounting to 102 seconds annually with respect to a fixed equinox.

458. *Period of the November Meteors.*—A comparison of the dates mentioned in Art. 454 shows that the grand displays recur after a cycle of about one third of a century, and that a grand display may occur on two consecutive years. A number greater than usual may be observed also for three or four consecutive years. Hence we must conclude that these meteors belong to a system of small bodies describing an elliptic orbit about the sun, and extending in the form of a stream along a considerable arc of that orbit. It is evident that the meteors can not make more than two complete revolutions in a year, for the major axis of an orbit which should be completed in one third of a year would not reach from the sun to the earth. Hence we conclude that in one year the group of meteors must describe either $2 \pm \frac{1}{33}$, or $1 \pm \frac{1}{33}$, or $\frac{1}{33}$ revolutions; that is, the periodic time must be either 180, 185, 354 or 376 days, or $33\frac{1}{4}$ years.

The motion of the node of a group of meteors describing an orbit about the sun in each of the preceding periods has been computed, and it is found that the motion corresponding to either of the first four mentioned periods would be entirely incompatible with the motion actually observed; but if the period be assumed $33\frac{1}{4}$ years, the computed motion of the node due to the action of the planets agrees almost exactly with the observed motion. This coincidence is regarded as demonstrating that the true period of the November meteors is $33\frac{1}{4}$ years.

458. *Elements of the November Meteors.*—Assuming the period as thus determined, and also the position of the radiant point shown by the observations, it is possible to compute the elements of the orbit. These elements are given in the first part of the following table:

	November Meteors.	Comet of 1866.
Period	33.25 years.	33.18 years.
Semi-axis major . . .	10.3402	10.3248
Eccentricity	0.9047	0.9054
Perihelion distance . .	0.9855	0.9765
Inclination	16° 46′	17° 18′
Longitude of node . .	51° 28′	51° 26′
Longitude of perihelion .	58° 19‛	60° 28′
Motion	Retrograde.	Retrograde.

Figure 96, p. 234, shows the form and dimensions of this orbit.

Fig. 96.

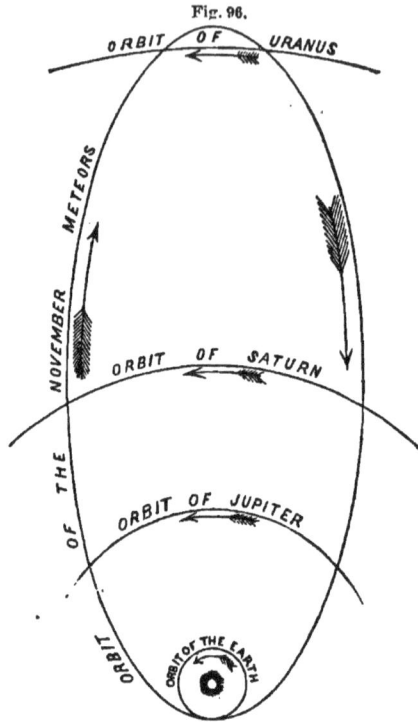

459. *First Comet of* 1866.—The elements of the first comet of
1866 bear a remarkable resemblance to those of the November
meteors.   These elements are given in the last column of the preceding table.   It is very improbable that so close a coincidence
should be accidental, and hence we seem authorized to conclude
that the comet of 1866 is *a very large meteor belonging to the November stream.*

460. *Dimensions of the November Stream.* — The November
stream of meteors is several years in passing its node.   The
length of the period during which extraordinary displays of meteors may occur is more than one year, and an unusual number
of shooting-stars, sufficient to attract attention, may be seen
through a period of at least 5 or 6 years.   Hence we conclude
that the length of the denser portion of the group, when at perihelion, is at least one fourth of the circumference of the orbit,

or one thousand millions of miles; while a large number of meteors extend still farther along the orbit. Since the shower of 1833 lasted two or three hours, the thickness of the ring at that point must have been the distance passed over by the earth in that time, multiplied by the sine of the inclination of the orbit, or about 50,000 miles. The comet of 1866 passed the path of the earth at a distance of six hundred thousand miles, which seems to imply that the breadth of the ring is much greater than its thickness.

461. *Conclusions.*—It thus appears to be pretty well established that the meteors of November are derived from a cosmical cloud, composed of very minute elements, each of which, before it encountered the earth, was moving in an elliptic path about the sun with a period of $33\frac{1}{4}$ years. This cloud has the form of an elliptic arc, the denser portion of which is at least 600 millions of miles in extent when near perihelion, and the rarer portion extends very much farther along the ellipse, while its thickness, where greatest, is over 50,000 miles. This cloud, although of immense extent, has very small density. It is computed that the mean distance of the individual elements of the group from each other when near perihelion is 30 or 40 miles; and although some of the meteors may have considerable size, their weight is doubtless very small. Hence the planets pass freely through the densest portion of this cloud without any sensible loss of motion.

462. *Division of Biela's Comet.*—Admitting that the November meteors have a period of $33\frac{1}{4}$ years, we find by computation that Biela's comet passed extremely near, and probably *through* the meteoric stream near the close of December, 1845. It has been conjectured that this collision may have produced the separation of this comet into two parts—a separation which was first noticed December 29th.

It is probable, however, that the density of the stream of meteors at this point was extremely small, so that this cause would seem inadequate to account for the division of Biela's comet.

463. *The Periodical Meteors of August.* — Another season at which meteors appear each year in unusual numbers occurs about the 10th of August. The periodicity of this display was established in 1837, since which time an extraordinary number of meteors has been uniformly observed each year, both in Europe and America, from the 6th to the 13th of the month, the greatest number being generally seen on the morning of the 10th. At the time of the maximum, the number of meteors visible is about three times as great as for the average of the entire month, and five times as great as for the average of the entire year.

The meteors of August, like those of November, seem also to emanate chiefly from a fixed point in the heavens. This point is in the constellation Perseus, being in R. A. 44°, and Dec. 56° N.

464. *Elements of the Orbit of the August Meteors.*—Assuming that the meteors radiated from the point just stated; that the orbit is a parabola, and that the earth crossed the centre of the group in 1866, Aug. 10.75, the elements given in the first part of the following table have been computed:

	August Meteors.	Third Comet of 1862.
Longitude of perihelion .	343° 28′	344° 41′ .
Longitude of node . .	138° 16′	137° 27′
Inclination	64° 3′	66° 26′
Perihelion distance . .	0.9643	0.9626
Period		121.5 years.
Motion	Retrograde.	Retrograde.

The elements of the third comet of 1862, given in the last column of the preceding table, bear a remarkable resemblance to those of the August meteors. The difference is no greater than can be accounted for by the want of precision in the data for computing the paths of the meteors. Hence we conclude that the great comet of 1862 was *one of the August meteors*, and probably one of the largest of that group.

465. *Dimensions of the August Stream.*—It is considered, then, highly probable that the August meteors describe a very large elliptic orbit about the sun, extending considerably beyond the orbit of Neptune. It is probable that the meteors are spread over the entire circumference of this orbit, but not in equal numbers. There are on record 63 remarkable displays of meteors which are considered to belong to this group, the earliest having occurred A.D. 811.   A comparison of these dates affords some indication of a maximum of brilliancy recurring at intervals of 108 years.

The earth, moving at the rate of 68,000 miles per hour, is at least seven days in passing entirely through the ring, which indicates that the thickness of the ring is more than eleven millions of miles.

The density of this stream of meteors is quite small, the mean distance of the individuals of the group from each other being computed to be more than a hundred miles.

466. *Origin of Meteoric Streams.*—Streams of meteors moving about the sun in orbits of vast extent may be supposed to have resulted from a nebulous mass, or cosmical cloud, acted upon by the attraction of the sun.   Let us suppose a cosmical cloud, consisting of very small meteors, to be drawn from stellar space by the attraction of the sun.   The individual particles of the cloud will move in elliptic orbits about the sun, but these ellipses will not be exactly equal to each other.   If the form of the cloud were at first spherical, its shape would be gradually changed, and it would ultimately be drawn out into a parabolic or elliptic arc, of which the sun is the focus.   If the orbit were an ellipse, the original form of the cloud would never be regained.   At each perihelion passage the length of the stream would be increased, and after a certain number of revolutions the cloud would become a continuous ring.   The stream would be at first periodic, but finally the flow would be constant.   If the primitive form of the group was not spherical, similar results would follow.   The meteors of November are supposed to belong to such a group, in which the ring is only partially formed, while the August meteors represent a group which has been transformed into a continuous ring.   Hence it is inferred that the November group is of comparatively recent formation.

467. *Other Periods of Shooting-stars.*—Besides the months of August and November, there are several other periods at which, either annually or occasionally, shooting-stars have been observed in unusual numbers.   Of these, the best established periods are shown in the following table, which also gives the radiant point from which the meteors seem chiefly to emanate.   These meteors are generally found to have a pretty definite radiant point, like the meteors of November and August.

Date of Display.		Radiant Point of the Meteors.			
Jan. 2 .	. A. R. 234°;	N.	Dec. 51°, near	ζ Cor. Borealis.	
April 20 .	" 277°;	"	35°,	"	α Lyræ.
July 28–29	" 304°;	"	40°,	"	γ Cygni.
Oct. 24 .	" 83°;	"	12°,	"	α Orionis.
Dec. 8–13	" 105°;	"	30°,	"	τ Geminorum.

The meteors which are seen on ordinary nights, and which do not show any marked uniformity of direction, have been called *sporadic.* It is, however, not improbable that meteors which at present are regarded as sporadic, may hereafter be proved to be periodical. It seems probable that shooting-stars, before they encounter the earth, form in the planetary spaces a multitude of currents or continuous rings, differing greatly in size and density, situated at various distances from the sun, and having all possible inclinations to the ecliptic. The unequal number of shooting-stars witnessed on different days of the year is the consequence of the unequal distribution of these meteoric streams throughout the planetary spaces.

SECTION II.

DETONATING METEORS.

468. *Detonating Meteors defined.* — Ordinary shooting-stars are not accompanied by any audible sound, although they are sometimes seen to break into pieces. Occasionally meteors of extraordinary brilliancy are succeeded by a loud detonation, or explosion, followed by a noise like that of musketry, or the discharge of cannon. These have been called detonating meteors.

469. *The New Jersey Meteor of November 15th,* 1859.—On the morning of November 15th, 1859, about 9½ o'clock, a remarkable meteor appeared in the heavens over the southern part of New Jersey. It was so brilliant that, although the sun was unclouded, and had an elevation of about 20° above the horizon, the flash attracted the attention of multitudes of persons as far north as Albany and Boston, and as far south as Fredericksburg, Virginia. Its apparent path was downward, inclined a few degrees to the west, and it left behind it a cloud of a rounded form like a puff of smoke. Soon after the flash there was heard a series of terrific explosions, which were compared to the discharge of a thousand cannon. These explosions were heard throughout Delaware and most of New Jersey. From a comparison of numerous observations, it was computed that the height of this meteor, when first seen, was over 60 miles, and when it exploded its height was 20 miles. The length of its visible path was more than 40 miles. It described this path in two seconds, so that its velocity relative to the earth was at least 20 miles per second. The column of smoke resulting from the explosions was a thousand feet in diameter, and several miles in length.

Comparing the motion of this meteor with that of the earth in its orbit, we find that its velocity relative to the sun was about 28 miles per second, which is the velocity belonging to a parabolic orbit. The lowest admissible estimate of its velocity would indicate that this meteor was moving about the sun in a very eccentric ellipse; the most probable velocity would indicate that its path was either a parabola or an hyperbola.

470. *The Tennessee Meteor of August 2d,* 1860.—On the 2d of August, 1860, about 10 P.M., a magnificent fire-ball was seen throughout the whole region from Pittsburg to New Orleans, and from Charleston to St. Louis, an area of nine hundred miles in diameter. It was described as equal in size to the full moon, and before its disappearance it broke into several fragments. A few minutes after its disappearance, there was heard throughout several counties of Kentucky and Tennessee a tremendous explosion like the sound of distant cannon.

From a comparison of a large number of observations, it has been computed that this meteor, when first seen, was about 82 miles above the earth's surface, and it exploded at an elevation of 28 miles. The length of its visible path was about 240 miles, and time of flight 8 seconds, showing a velocity relative to the earth of 30 miles per second. It is hence computed that its velocity relative to the sun was 24 miles per second.

471. *Number, Velocity, etc.*—Examples of detonating meteors similar to the preceding are of yearly occurrence, and if every case was duly reported, they would probably be found to be of daily and perhaps hourly occurrence. The number of detonating

meteors found recorded in scientific journals is over 800. Their
average height at the first instant of apparition is 92 miles, and at
the instant of vanishing is 32 miles. Their average velocity relative to the earth is estimated at 19 miles per second.

472. *Multiple Nuclei, etc.*—Sometimes the head of a meteor appears divided, consisting of two or more brilliant bodies in the
form of elongated drops, each followed by a tail of fiery appearance. In a few cases as many as a dozen heads have been counted, but generally these secondary heads follow the principal body
of light so closely that they give to the meteor an elongated appearance, which has been sometimes compared to a child's kite, a
pear, a fish, etc.

The track of the meteor is often marked by a permanent streak,
which sometimes continues visible for many minutes. This streak
gradually changes its shape and position, like a cloud moved by
the wind, sometimes assuming a serpentine form, sometimes bend-
ing up like a crescent or a horse-shoe, and drifting with a velocity
of more than 100 miles per hour.

473. *Periodicity of Detonating Meteors.*—An unusual number of
detonating meteors has been seen about the time of the grand
meteoric display of November 13th; also about the time of the
grand display of August 10th; and also December 8th–13th.
Moreover, several detonating meteors have been recorded January 2d and April 20th. This coincidence in the times of unusual display of detonating meteors and of ordinary shooting-stars,
taken in connection with the results obtained respecting their .
paths and velocities, leads us to infer that both belong to the same
class of bodies, and that they do not probably differ much from
each other except in size and density. We conclude, then, that
detonating meteors are small bodies which revolve about the sun
in orbits which are generally ellipses of considerable eccentricity,
but perhaps sometimes parabolas or even hyperbolas. They are
bodies of considerable density, and the noise which succeeds their
appearance is probably in great part due to the collapse of the
air rushing into the vacuum which is left behind the advancing
meteor. No audible sound proceeds from ordinary shooting-
stars, because they are bodies of small size or of feeble density,
and are generally dissipated or consumed while yet at an elevation of 50 miles above the earth's surface.

SECTION III.

AEROLITES.

474. *Aerolites described.*—There is no evidence that any deposit from ordinary shooting-stars ever reaches the earth's surface, but occasionally solid substances descend to the earth from beyond the earth's atmosphere. The fragments generally penetrate a foot or more into the earth, and if picked up soon after their fall are found to be warm, and sometimes even hot. These small bodies are called *aerolites*. They are called *meteoric stones* when they present a stony appearance, or *meteoric iron* when they are almost entirely metallic.

Although numerous instances of the fall of aerolites had been recorded from the earliest historic times, and especially during the last century, these accounts were received by many scientific men with incredulity. But during the present century these cases have been so numerous, and they have been witnessed by so many observers, that we can no longer doubt that stones have fallen to the earth from beyond the earth's atmosphere.

475. *The Weston, Connecticut, Aerolite.*—On the morning of December 14th, 1807, a meteor of great brilliancy was seen moving through the atmosphere over the town of Weston, Connecticut. Its apparent diameter was about one half that of the full moon; and soon after its disappearance there were heard by those nearly under the place of disappearance three loud explosions like those of a cannon, followed by a quick succession of smaller reports. Immediately after the explosions, one observer heard a sound like that occasioned by the fall of a heavy body, and, upon examination, found that a stone had fallen upon a rock near his house, and was broken into small fragments. The fragments were still warm, and together were estimated to weigh about twenty pounds.

In another place, about five miles from the former, a fresh hole was found in the turf, and at the bottom of the hole, at the depth of two feet, was found a stone weighing thirty-five pounds. In the neighborhood was found a third stone weighing about ten pounds, a fourth weighing thirteen pounds, a fifth weighing twenty pounds, and a sixth weighing thirty-six pounds. At a spot

Q

about four miles distant from the preceding, a large mass of stones, estimated to weigh 200 pounds, fell upon a rock, and was broken into minute fragments. It was estimated that the entire weight of all the fragments was at least 300 pounds.

The specimens from all these localities were quite similar, and their specific gravity varied from 3.3 to 3.6. Their composition was nearly one half silex, about one third oxyd of iron, and one sixth magnesia, with a little nickel and sulphur.

The same meteor was extensively seen as far north as Vermont, and as far south as New Jersey. The length of its visible path exceeded 100 miles, and it moved from northwest to southeast, its path being inclined downward about 30° to the horizon, and when it exploded its elevation was only about eight miles. The time of flight was probably between five and ten seconds. Hence the velocity relative to the earth was about fifteen miles per second.

476. *The Guernsey, Ohio, Aerolite.*—On the first of May, 1860, about half an hour after noon, an aerolite exploded over Guernsey County, Ohio. A great number of distinct detonations were heard, like the firing of a cannon, after which the sounds became blended together, and were compared to the roar of a railway train. The elevation of this meteor above the earth's surface was computed at forty-one miles, and its path was nearly horizontal. The entire weight of all the fragments which descended from this meteor was estimated at 700 pounds. Their specific gravity was 3.54, and their composition very similar to that of the Weston meteor.

477. *The Braunau, Bohemia, Aerolite.*—On the 14th of July, 1847, about four o'clock in the morning, at Braunau, in Bohemia, there were heard two heavy explosions, which followed each other in quick succession. Two streams of fire were seen to descend to the earth, and, upon examination, a fresh hole three feet deep was found in the earth, and at the bottom of the hole a mass of iron, which for six hours after the fall continued so hot that it could not be held in the hand. This mass weighed forty-two pounds, and is preserved in the cabinet at Vienna. Another mass, weighing thirty pounds, fell upon a roof, and broke through large pieces of timber.

The specific gravity of this meteor was 7.71.   Its composition was ninety-two per cent. of iron, and five per cent. of nickel, with a small quantity of cobalt, arsenic, etc.

478. *The Orgueil, France, Aerolite.* — On the evening of May 14th, 1864, a very bright fire-ball was seen in France, throughout the whole region from Paris to the Pyrenees.   Loud detonations were heard in the neighborhood of Montauban, and a large number of stones fell near the village of Orgueil.   The passage of the meteor was witnessed by a large number of intelligent observers. It was first seen at an altitude greater than fifty-five miles; it exploded at an altitude of about twenty miles; and it was descending in a line inclined 20° or 25° to the horizon.   The length of its visible path was 112 miles; and the time of flight was estimated at five or six seconds, indicating a velocity of not less than fifteen or twenty miles per second.   The stones were hot when they were first picked up.   Their specific gravity was 2.567.

479. *Number of Aerolites.*—There are eighteen well-authenticated cases in which aerolites have fallen in the United States during the last sixty years, and their aggregate weight is 1250 pounds.   The entire number of known aerolites, the date of whose fall is well determined, is 261.   There are also on record seventy-four cases of aerolites in which the day and month are not given, and sometimes even the year is uncertain.   Besides these there have been found eighty-six masses, which, from their peculiar composition, are believed to be aerolites, although the date of their fall is unknown.   The weight of these masses varies from a few pounds to several tons.   The entire number of aerolites of which we have any knowledge is therefore about 420.

The actual number of aerolites which have reached the earth must have been far greater than this.   Many must have fallen upon the ocean, or upon uninhabited lands where they were unobserved.   During the past fifty years the fall of 115 aerolites has been recorded.   If we suppose aerolites to have fallen over the entire globe at the same rate as has been observed over the more populous portions of Europe and America, we should have an average of over 300 annually.   Now we can not suppose that even in Europe more than half the whole number are actually seen to fall; hence we conclude that more than 600 aerolites fall annually on various parts of the earth's surface. If we suppose their average weight to equal that of those which have fallen in the United States, we should have for the entire globe eighteen tons of aerolites annually. See Tables XXXV. and XXXVI.

480. *Chemical Composition of Aerolites.*—Aerolites are composed of the same elementary substances as occur in terrestrial minerals, not a single new element having been found in their analysis. Of the sixty-three elements now admitted by chemists, the following twenty or twenty-two have been found in aerolites.

Metals.		Metalloids.
1. Aluminium.	9. Manganese.	1. Carbon.
2. Calcium.	10. Nickel.	2. Oxygen.
3. Chromium.	11. Potassium.	3. Phosphorus.
4. Cobalt.	12. Sodium.	4. Silicium.
5. Copper.	13. Strontium.	5. Sulphur.
6. Iron.	14 Tin.	6. Arsenic?
7. Lithium.	15. Titanium.	7. Chlorine?
8. Magnesium.		

Aerolites differ greatly in the proportions of these ingredients. Some of them contain ninety-six per cent. of iron, while others contain less than one per cent. Some contain eighteen per cent. of nickel, and others less than one per cent. On the contrary, others consist mostly of silica, magnesia, lime, etc. It is common, therefore, to divide aerolites into two groups, viz., meteoric iron and meteoric stones.

The specific gravity of aerolites varies from 1.70 (that of March 15th, 1806, at Alais, France) to 7.8 (that of May 26th, 1751, at Agram, in Austria).

481. *Peculiarities of Aerolites.*—While aerolites contain no elements but such as are found in terrestrial minerals, their appearance is quite peculiar, and the grouping of the elements, that is, the compounds formed by them, are so peculiar as to enable us by chemical analysis to distinguish an aerolite from any terrestrial substance.

Iron ores are very abundant in nature, but iron in the metallic state is exceedingly rare in nature. Now aerolites invariably contain metallic iron, sometimes ninety to ninety-six per cent.

This iron is perfectly malleable, and may be readily worked into cutting instruments. This meteoric iron *always* contains a certain amount of nickel, generally eight or ten per cent., with small quantities of cobalt, copper, tin, and chrome. This composition *has never been found in any terrestrial mineral*. Moreover, when the fragments of meteoric iron which are dispersed through those aerolites which are mostly earthy are extracted and submitted to analysis, they show the same composition, viz., about ninety of iron, with eight or ten of nickel, etc.

Many of the other constituents of aerolites are similar to those which are found in volcanic rocks, such as olivine (a silicate of magnesia), magnetic pyrites, chrome-iron, etc.

All aerolites, without exception, contain a substance called *schreibersite*, though often in very small quantities. This substance is a compound of iron, nickel, and phosphorus, and *has never been found except in aerolites*. Fig. 97 represents an iron meteor found near Lockport, New York, in 1818.

Fig. 97.

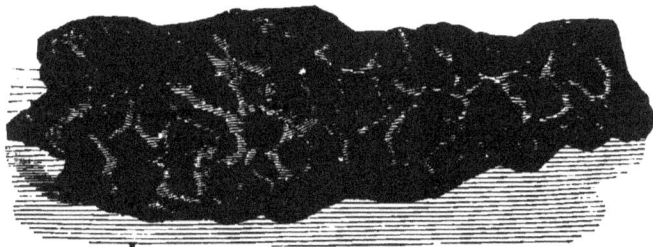

482. *Widmannstäten Figures.*—Meteoric iron possesses a highly crystalline structure. If the surface be carefully polished, and the mass be heated to a straw-yellow, after cooling, the surface will be covered with groups of regular triangles formed by lines nearly parallel to each other, intersected by others at angles of sixty degrees. These figures were first discovered by an Austrian iron - master, Widmannstäten, in the year 1808, and they have received the name of their discoverer.

It was afterward discovered that the same figures could be developed by the use of acids. For this purpose, nitric acid is diluted with an equal volume of water, and the iron, having been previously cut and polished, is placed in the solution, the parts not required to be acted upon being coated with asphaltum.

After five or six minutes the iron is taken out of the acid, carefully washed and dried. Figure 98 shows the crystalline structure of the meteoric iron of Elbogen, preserved in the cabinet of Vienna.

Fig. 98.

Ordinary iron will not exhibit these Widmannstäten figures, but iron melted directly out of some volcanic rocks does exhibit them.

483. *Periodicity of Aerolites.*—The falls of aerolites exhibit some indications of periodicity, and these periods correspond with those of ordinary shooting-stars. There are on record éleven cases in which aerolites have been seen to fall near the time of the annual display of the August meteors, Art. 463; that is, four per cent. of all the recorded aerolite falls have occurred within three days of the maximum display of August meteors. This number is more than double that which we should expect if aerolites and shooting-stars had no connection with each other.

There are on record seven cases in which aerolites have fallen between December 7th and 13th, which is also a period of unusual display of shooting-stars, Art. 467; and there are also three cases in which aerolites have fallen from November 11th to 13th. These numbers are greater than should be expected if shooting-stars and aerolites were entirely independent of each other.

It is not probable that such a coincidence of dates is accidental, and hence we are led to conclude that aerolites form portions of the nebulous rings or groups from which shooting-stars are derived.

483. *Are Aerolites formed in our Atmosphere?*—Various hypotheses have been proposed to account for the origin of aerolites. It has been conjectured that they are formed in our atmosphere like rain or hail. This supposition is inadmissible, because, allowing the aerolite to be once formed, there is no known force which could impel it in a direction nearly horizontal with a velocity of several miles per second.

484. *Have Aerolites been ejected from Terrestrial Volcanoes?*—It has been conjectured that aerolites are masses ejected from terrestrial volcanoes. This supposition is inadmissible, because the greatest velocity with which stones have ever been ejected from volcanoes is less than two miles per second, and the direction of this motion must be nearly vertical, while aerolites frequently move in a direction nearly horizontal, and with a velocity of several miles per second. This argument is unanswerable, and therefore it is superfluous to add that the composition of aerolites is different from that of any known terrestrial mineral.

485. *Have Aerolites been ejected from Lunar Volcanoes?*—It has been conjectured that aerolites have been ejected from volcanoes in the moon with a velocity sufficient to carry them out of the sphere of the moon's attraction into that of the earth's attraction. It has been computed that a velocity of projection of 8000 feet per second would be sufficient to produce such an effect.

The following are some of the objections to this hypothesis:

1. In order that a body projected from the moon may reach the earth's surface, it must describe about the earth a conic section whose distance at perigee is less than the earth's radius. Hence there are limits to the direction in which the aerolite must have left the moon, and also to the force with which it must have been projected. If a body was projected from near the moon's centre, or from its eastern hemisphere, since it would retain the moon's orbital velocity, its resulting velocity would be such that its perigee distance would exceed 4000 miles. If the body was projected with a small force, it would not get beyond the sphere of the

moon's attraction; and if the velocity was too great, the perigee distance would exceed 4000 miles. It has been computed that a change of $\frac{1}{180}$ part in the force of projection would cause a change of more than 4000 miles in the perigee distance, and for a given force of projection a change of $\frac{1}{180}$ part in the mass of the body would produce a like effect.

Hence it has been estimated that if an indefinite number of bodies, having different masses, were expelled from the moon in all directions and with different velocities, not one in a million could reach the earth. But it is computed that 600 aerolites fall to the earth annually, Art. 479. Hence the lunar hypothesis requires us to conclude that more than 600 millions of aerolites are annually expelled from the moon. But the lunar volcanoes are to all appearance nearly, if not entirely, extinct; and although the moon has long been carefully watched with the most powerful telescopes, in only one or two instances have astronomers suspected that they had discovered any indications of change. We can not, therefore, admit that lunar volcanoes have ejected rocks in such quantities as to account for the known aerolites.

2. The observed velocities of some aerolites are incompatible with the theory that they are satellites of the earth. In order that a body may revolve around the earth, its velocity must not be less than 5 miles, nor greater than 7 miles per second. If the velocity were less than 5, the body would fall to the earth; and if the velocity was greater than 7, the body would recede from the earth, never to return. Now the velocity of the Orgueil meteor, Art. 478, certainly exceeded 7 miles per second, and therefore it was not a satellite to the earth. There are but few cases in which the velocity of aerolites has been even rudely determined; but detonating meteors seem to have the same origin as aerolites, and the average velocity of detonating meteors is certainly greater than 7 miles per second.

3. Aerolites appear to be subject to a periodicity depending upon the season of the year, which shows that they are satellites of the sun and not of the earth.

Although, then, we can not pronounce it impossible that a small body projected from a lunar volcano may occasionally have fallen to the earth, it is certain that aerolites generally can not have had this origin, and there is no reason to suppose that any aerolite has ever been derived from this source.

486. *Conclusions.*—A comparison of all the facts which are known respecting shooting-stars, detonating meteors, and aerolites leads to the conclusion that they are all minute bodies revolving like the comets in orbits about the sun, and are encountered by the earth in its orbital motion. The visible path of aerolites is somewhat nearer to the earth's surface than that of ordinary shooting-stars, a result which may be ascribed to their greater density. It is probable, also, that their velocity is somewhat smaller, a result which may be due to their descending into an atmosphere of greater density, which causes, therefore, greater resistance.

These three classes of bodies exhibit alternate periods of maximum and minimum abundance, and the times of maximum for the several classes correspond somewhat with each other, indicating that these bodies are collected in groups, and the three classes of bodies are grouped in a somewhat similar manner. The August meteors move in orbits which require more than a century to complete, and comprehend bodies differing greatly in size and probably also in density. Their magnitudes range from comets whose diameter is perhaps 100,000 miles to minute atoms which, in a single second, are dissipated by the heat resulting from their collision with our atmosphere. Their density ranges from that of metallic iron to earthy bodies having but feeble cohesion, which are dissipated into fine dust by the heat of collision with our atmosphere; and it is possible that the rarest of them may consist of solid or liquid matter in a state of minute subdivision, like a cloud of dust or smoke.

The periodic meteors of November probably comprehend bodies having an equal range of magnitude, and perhaps also of density.

TABLE I.—MILLIMETRES CONVERTED INTO INCHES. 251

TABLE I.

TO CONVERT MILLIMETRES INTO ENGLISH INCHES.

Millimetres.	Inches.	Millimetres.	Inches.	Millimetres.	Inches.	Millimetres.	Inches.	Millimetres.	Inches.	Millimetres.	Inches.
1	0.039	50	1.969	500	19.685	689	27.126	734	28.898	779	30.670
2	.079	60	2.362	510	20.079	690	.166	735	.938	780	.709
3	.118	70	2.756	520	20.473	691	.205	736	.977	781	.749
4	.157	80	3.150	530	20.867	692	.245	737	29.016	782	.788
5	.197	90	3.543	540	21.260	693	.284	738	.056	783	.827
6	.236	100	3.937	550	21.654	694	.323	739	.095	784	.867
7	.276	110	4.331	560	22.048	695	.363	740	.134	785	.906
8	.315	120	4.724	570	22.441	696	.402	741	.174	786	.945
9	.354	130	5.118	580	22.835	697	.441	742	.213	787	.985
10	.394	140	5.512	590	23.229	698	.481	743	.252	788	31.024
11	.433	150	5.906	600	23.622	699	.520	744	.292	789	.064
12	.472	160	6.299	610	24.016	700	.560	745	.331	790	.103
13	.512	170	6.693	620	24.410	701	.599	746	.371	791	.142
14	.551	180	7.087	630	24.804	702	.638	747	.410	792	.182
15	.591	190	7.480	640	25.197	703	.678	748	.449	793	.221
16	.630	200	7.874	650	25.591	704	.717	749	.489	794	.260
17	.669	210	8.268	660	25.985	705	.756	750	.528	795	.300
18	.709	220	8.662	661	26.024	706	.796	751	.567	796	.339
19	.748	230	9.055	662	.063	707	.835	752	.607	797	.379
20	.787	240	9.449	663	.103	708	.875	753	.646	798	.418
21	.827	250	9.843	664	.142	709	.914	754	.686	799	.457
22	.866	260	10.236	665	.182	710	.953	755	.725	800	.497
23	.905	270	10.630	666	.221	711	.993	756	.764	810	.890
24	.945	280	11.024	667	.260	712	28.032	757	.804	820	32.284
25	.984	290	11.418	668	.300	713	.071	758	.843	830	.678
26	1.024	300	11.811	669	.339	714	.111	759	.882	840	33.072
27	.063	310	12.205	670	.378	715	.150	760	.922	850	.465
28	.102	320	12.599	671	.418	716	.189	761	.961	860	.859
29	.142	330	12.992	672	.457	717	.229	762	30.001	870	34.253
30	.181	340	13.386	673	.497	718	.268	763	.040	880	.646
31	.220	350	13.780	674	.536	719	.308	764	.079	890	35.040
32	.260	360	14.173	675	.575	720	.347	765	.119	900	.434
33	.299	370	14.567	676	.615	721	.386	766	.158	Proportional Parts.	
34	.339	380	14.961	677	.654	722	.426	767	.197		
35	.378	390	15.355	678	.693	723	.465	768	.237	Mill.	Inches.
36	.417	400	15.748	679	.733	724	.504	769	.276	0.1	0.004
37	.457	410	16.142	680	.772	725	.544	770	.316	0.2	0.008
38	.496	420	16.536	681	.812	726	.583	771	.355	0.3	0.012
39	.535	430	16.929	682	.851	727	.623	772	.394	0.4	0.016
40	.575	440	17.323	683	.890	728	.662	773	.434	0.5	0.020
41	.614	450	17.717	684	.930	729	.701	774	.473	0.6	0.024
42	.654	460	18.111	685	.969	730	.741	775	.512	0.7	0.028
43	.693	470	18.504	686	27.008	731	.780	776	.552	0.8	0.031
44	.732	480	18.898	687	.048	732	.819	777	.591	0.9	0.035
45	.772	490	19.292	688	.087	733	.859	778	.630	1.0	0.039

One millimetre equals 0.03937079 English inch.

TABLE II.

TO CONVERT METRES INTO ENGLISH FEET.

Metres.	Feet.	Metres.	Feet.	Metres.	Feet.	Metres.	Feet.	Metres.	Feet.	Metres.	Feet.
1	3.28	46	150.92	91	298.56	136	446.20	181	593.84	226	741.48
2	6.56	47	154.20	92	301.84	137	449.48	182	597.12	227	744.76
3	9.84	48	157.48	93	305.12	138	452.76	183	600.40	228	748.05
4	13.12	49	160.76	94	308.40	139	456.04	184	603.69	229	751.33
5	16.40	50	164.04	95	311.69	140	459.33	185	606.97	230	754.61
6	19.69	51	167.33	96	314.97	141	462.61	186	610.25	231	757.89
7	22.97	52	170.61	97	318.25	142	465.89	187	613.53	232	761.17
8	26.25	53	173.89	98	321.53	143	469.17	188	616.81	233	764.45
9	29.53	54	177.17	99	324.81	144	472.45	189	620.09	234	767.73
10	32.81	55	180.45	100	328.09	145	475.73	190	623.37	235	771.01
11	36.09	56	183.73	101	331.37	146	479.01	191	626.65	236	774.29
12	39.37	57	187.01	102	334.65	147	482.29	192	629.93	237	777.57
13	42.65	58	190.29	103	337.93	148	485.57	193	633.21	238	780.85
14	45.93	59	193.57	104	341.21	149	488.85	194	636.49	239	784.13
15	49.21	60	196.85	105	344.49	150	492.13	195	639.78	240	787.42
16	52.49	61	200.13	106	347.78	151	495.42	196	643.06	241	790.70
17	55.78	62	203.42	107	351.06	152	498.70	197	646.34	242	793.98
18	59.06	63	206.70	108	354.34	153	501.98	198	649.62	243	797.26
19	62.34	64	209.98	109	357.62	154	505.26	199	652.90	244	800.54
20	65.62	65	213.26	110	360.90	155	508.54	200	656.18	245	803.82
21	68.90	66	216.54	111	364.18	156	511.82	201	659.46	246	807.10
22	72.18	67	219.82	112	367.46	157	515.10	202	662.74	247	810.38
23	75.46	68	223.10	113	370.74	158	518.38	203	666.02	248	813.66
24	78.74	69	226.38	114	374.02	159	521.66	204	669.30	249	816.94
25	82.02	70	229.66	115	377.30	160	524.94	205	672.58	250	820.22
26	85.30	71	232.94	116	380.58	161	528.22	206	675.87	251	823.51
27	88.58	72	236.22	117	383.87	162	531.51	207	679.15	252	826.79
28	91.87	73	239.51	118	387.15	163	534.79	208	682.43	253	830.07
29	95.15	74	242.79	119	390.43	164	538.07	209	685.71	254	833.35
30	98.43	75	246.07	120	393.71	165	541.35	210	688.99	255	836.63
31	101.71	76	249.35	121	396.99	166	544.63	211	692.27	256	839.91
32	104.99	77	252.63	122	400.27	167	547.91	212	695.55	257	843.19
33	108.27	78	255.91	123	403.55	168	551.19	213	698.83		
34	111.55	79	259.19	124	406.83	169	554.47	214	702.11		
35	114.83	80	262.47	125	410.11	170	557.75	215	705.39		
36	118.11	81	265.75	126	413.39	171	561.03	216	708.67		
37	121.39	82	269.03	127	416.67	172	564.31	217	711.96		
38	124.67	83	272.31	128	419.96	173	567.60	218	715.24		
39	127.96	84	275.60	129	423.24	174	570.88	219	718.52		
40	131.24	85	278.88	130	426.52	175	574.16	220	721.80		
41	134.52	86	282.16	131	429.80	176	577.44	221	725.08		
42	137.80	87	285.44	132	433.08	177	580.72	222	728.36		
43	141.08	88	288.72	133	436.36	178	584.00	223	731.64		
44	144.36	89	292.00	134	439.64	179	587.28	224	734.92		
45	147.64	90	295.28	135	442.92	180	590.56	225	738.20		

Proportional Parts.

Met.	Feet.
0.1	0.33
0.2	0.66
0.3	0.98
0.4	1.31
0.5	1.64
0.6	1.97
0.7	2.30
0.8	2.62
0.9	2.95
1.0	3.28

One metre equals 3.2808992 English feet.

TABLE III.—KILOMETRES CONVERTED INTO MILES. 253

TABLE III.

TO CONVERT KILOMETRES INTO ENGLISH MILES.

Kilometres.	Miles.	Kilometres.	Miles.	Kilometres.	Miles.	Kilometres.	Miles.	Kilometres.	Miles.	Kilometres.	Miles.
1	0.621	46	28.584	91	56.546	136	84.508	181	112.470	226	140.432
2	1.243	47	29.205	92	57.167	137	85.129	182	113.092	227	141.054
3	1.864	48	29.826	93	57.789	138	85.751	183	113.713	228	141.675
4	2.486	49	30.448	94	58.410	139	86.372	184	114.334	229	142.297
5	3.107	50	31.069	95	59.031	140	86.994	185	114.956	230	142.918
6	3.728	51	31.691	96	59.653	141	87.615	186	115.577	231	143.539
7	4.350	52	32.312	97	60.274	142	88.236	187	116.198	232	144.161
8	4.971	53	32.933	98	60.895	143	88.858	188	116.820	233	144.782
9	5.592	54	33.555	99	61.517	144	89.479	189	117.441	234	145.403
10	6.214	55	34.176	100	62.138	145	90.100	190	118.063	235	146.025
11	6.835	56	34.797	101	62.760	146	90.722	191	118.684	236	146.646
12	7.457	57	35.419	102	63.381	147	91.343	192	119.305	237	147.268
13	8.078	58	36.040	103	64.002	148	91.965	193	119.927	238	147.889
14	8.699	59	36.662	104	64.624	149	92.586	194	120.548	239	148.510
15	9.321	60	37.283	105	65.245	150	93.207	195	121.170	240	149.132
16	9.942	61	37.904	106	65.867	151	93.829	196	121.791	241	149.753
17	10.563	62	38.526	107	66.488	152	94.450	197	122.412	242	150.375
18	11.185	63	39.147	108	67.109	153	95.071	198	123.034	243	150.996
19	11.806	64	39.768	109	67.731	154	95.693	199	123.655	244	151.617
20	12.428	65	40.390	110	68.352	155	96.314	200	124.276	245	152.239
21	13.049	66	41.011	111	68.973	156	96.936	201	124.898	246	152.860
22	13.670	67	41.633	112	69.595	157	97.557	202	125.519	247	153.481
23	14.292	68	42.254	113	70.216	158	98.178	203	126.141	248	154.103
24	14.913	69	42.875	114	70.838	159	98.800	204	126.762	249	154.724
25	15.535	70	43.497	115	71.459	160	99.421	205	127.383	250	155.346
26	16.156	71	44.118	116	72.080	161	100.043	206	128.005	251	155.967
27	16.777	72	44.740	117	72.702	162	100.664	207	128.626	252	156.588
28	17.399	73	45.361	118	73.323	163	101.285	208	129.248	253	157.210
29	18.020	74	45.982	119	73.944	164	101.907	209	129.869	254	157.831
30	18.641	75	46.604	120	74.566	165	102.528	210	130.490	255	158.452
31	19.263	76	47.225	121	75.187	166	103.149	211	131.112	256	159.074
32	19.884	77	47.846	122	75.809	167	103.771	212	131.733	257	159.695
33	20.506	78	48.468	123	76.430	168	104.392	213	132.354		
34	21.127	79	49.089	124	77.051	169	105.014	214	132.976		
35	21.748	80	49.711	125	77.673	170	105.635	215	133.597		
36	22.370	81	50.332	126	78.294	171	106.256	216	134.219		
37	22.991	82	50.953	127	78.916	172	106.878	217	134.840		
38	23.613	83	51.575	128	79.537	173	107.499	218	135.461		
39	24.234	84	52.196	129	80.158	174	108.121	219	136.083		
40	24.855	85	52.818	130	80.780	175	108.742	220	136.704		
41	25.477	86	53.439	131	81.401	176	109.363	221	137.326		
42	26.098	87	54.060	132	82.022	177	109.985	222	137.947		
43	26.719	88	54.682	133	82.644	178	110.606	223	138.568		
44	27.341	89	55.303	134	83.265	179	111.227	224	139.190		
45	27.962	90	55.924	135	83.887	180	111.849	225	139.811		

Proportional Parts.

Kil.	Miles.
0.1	0.062
0.2	0.124
0.3	0.186
0.4	0.249
0.5	0.311
0.6	0.373
0.7	0.435
0.8	0.497
0.9	0.559
1.0	0.621

One kilometre equals 0.6213824 English mile.

TABLE IV.

TO CONVERT FRENCH FEET INTO ENGLISH FEET.

Fr.	English.	Fr.	English.	Fr.	English.	Fr.	English.	Fr.	English.	Fr.	English.
1	1.066	46	49.025	91	96.985	136	144.944	181	192.903	226	240.863
2	2.132	47	50.091	92	98.050	137	146.010	182	193.969	227	241.929
3	3.197	48	51.157	93	99.116	138	147.076	183	195.035	228	242.995
4	4.263	49	52.222	94	100.182	139	148.141	184	196.101	229	244.060
5	5.329	50	53.288	95	101.248	140	149.207	185	197.167	230	245.126
6	6.395	51	54.354	96	102.314	141	150.273	186	198.232	231	246.192
7	7.460	52	55.420	97	103.379	142	151.339	187	199.298	232	247.258
8	8.526	53	56.486	98	104.445	143	152.404	188	200.364	233	248.323
9	9.592	54	57.551	99	105.511	144	153.470	189	201.430	234	249.389
10	10.658	55	58.617	100	106.576	145	154.536	190	202.495	235	250.455
11	11.723	56	59.683	101	107.642	146	155.602	191	203.561	236	251.521
12	12.789	57	60.749	102	108.708	147	156.667	192	204.627	237	252.586
13	13.855	58	61.814	103	109.774	148	157.733	193	205.693	238	253.652
14	14.921	59	62.880	104	110.840	149	158.799	194	206.758	239	254.718
15	15.986	60	63.946	105	111.905	150	159.865	195	207.824	240	255.784
16	17.052	61	65.012	106	112.971	151	160.931	196	208.890	241	256.849
17	18.118	62	66.077	107	114.037	152	161.996	197	209.956	242	257.915
18	19.184	63	67.143	108	115.103	153	163.062	198	211.021	243	258.981
19	20.250	64	68.209	109	116.168	154	164.128	199	212.087	244	260.047
20	21.315	65	69.275	110	117.234	155	165.194	200	213.153	245	261.113
21	22.381	66	70.340	111	118.300	156	166.259	201	214.219	246	262.178
22	23.447	67	71.407	112	119.366	157	167.325	202	215.285	247	263.244
23	24.513	68	72.472	113	120.431	158	168.391	203	216.350	248	264.310
24	25.578	69	73.538	114	121.497	159	169.457	204	217.416	249	265.376
25	26.644	70	74.604	115	122.563	160	170.522	205	218.482	250	266.441
26	27.710	71	75.669	116	123.629	161	171.588	206	219.548	251	267.507
27	28.776	72	76.735	117	124.695	162	172.654	207	220.613	252	268.573
28	29.841	73	77.801	118	125.760	163	173.720	208	221.679	253	269.639
29	30.907	74	78.867	119	126.826	164	174.785	209	222.745	254	270.704
30	31.973	75	79.932	120	127.892	165	175.851	210	223.811	255	271.770
31	33.039	76	80.998	121	128.958	166	176.917	211	224.877	256	272.836
32	34.104	77	82.064	122	130.023	167	177.983	212	225.942	257	273.902
33	35.170	78	83.130	123	131.089	168	179.049	213	227.008	Proportional Parts.	
34	36.236	79	84.195	124	132.155	169	180.114	214	228.074	Fr.	English.
35	37.302	80	85.261	125	133.221	170	181.180	215	229.140	0.1	0.107
36	38.368	81	86.327	126	134.286	171	182.246	216	230.205	0.2	0.213
37	39.433	82	87.393	127	135.352	172	183.312	217	231.271	0.3	0.320
38	40.499	83	88.458	128	136.418	173	184.377	218	232.337	0.4	0.426
39	41.565	84	89.524	129	137.484	174	185.443	219	233.403	0.5	0.533
40	42.631	85	90.590	130	138.549	175	186.509	220	234.468	0.6	0.639
41	43.696	86	91.656	131	139.615	176	187.575	221	235.534	0.7	0.746
42	44.762	87	92.722	132	140.681	177	188.640	222	236.600	0.8	0.853
43	45.828	88	93.787	133	141.747	178	189.706	223	237.666	0.9	0.959
44	46.894	89	94.853	134	142.813	179	190.772	224	238.731	1.0	1.066
45	47.959	90	95.919	135	143.878	180	191.838	225	239.797		

One French foot equals 1.065765 English foot.

TABLE V.—CENTESIMAL AND FAHRENHEIT THERMOMETERS. 255

TABLE V.

TO COMPARE THE CENTESIMAL THERMOMETER WITH FAHRENHEIT'S.

Centes.	Fahren.	Centes.	Fahren.	Centes.	Fahren.	Centes.	Fahren.	Proportional Parts.	
o	o	o	o	o	o	o	o	Centes.	Fahren.
								o	o
100	212.0	50	122.0	25	77.0	0	+32.0	0.1	0.18
99	210.2	49.5	121.1	24.5	76.1	— 1	30.2	0.2	0.36
98	208.4	49	120.2	24	75.2	— 2	28.4	0.3	0.54
97	206.6	48.5	119.3	23.5	74.3	— 3	26.6	0.4	0.72
96	204.8	48	118.4	23	73.4	— 4	24.8	0.5	0.90
95	203.0	47.5	117.5	22.5	72.5	— 5	23.0	0.6	1.08
94	201.2	47	116.6	22	71.6	— 6	21.2	0.7	1.26
93	199.4	46.5	115.7	21.5	70.7	— 7	19.4	0.8	1.44
92	197.6	46	114.8	21	69.8	— 8	17.6	0.9	1.62
91	195.8	45.5	113.9	20.5	68.9	— 9	15.8	1.0	1.80
90	194.0	45	113.0	20	68.0	—10	14.0		
89	192.2	44.5	112.1	19.5	67.1	—11	12.2		
88	190.4	44	111.2	19	66.2	—12	10.4		
87	188.6	43.5	110.3	18.5	65.3	—13	8.6		
86	186.8	43	109.4	18	64.4	—14	6.8		
85	185.0	42.5	108.5	17.5	63.5	—15	5.0		
84	183.2	42	107.6	17	62.6	—16	3.2		
83	181.4	41.5	106.7	16.5	61.7	—17	+ 1.4		
82	179.6	41	105.8	16	60.8	—18	— 0.4		
81	177.8	40.5	104.9	15.5	59.9	—19	— 2.2		
80	176.0	40	104.0	15	59.0	—20	— 4.0		
79	174.2	39.5	103.1	14.5	58.1	—21	— 5.8		
78	172.4	39	102.2	14	57.2	—22	— 7.6		
77	170.6	38.5	101.3	13.5	56.3	—23	— 9.4		
76	168.8	38	100.4	13	55.4	—24	—11.2		
75	167.0	37.5	99.5	12.5	54.5	—25	—13.0		
74	165.2	37	98.6	12	53.6	—26	—14.8		
73	163.4	36.5	97.7	11.5	52.7	—27	—16.6		
72	161.6	36	96.8	11	51.8	—28	—18.4		
71	159.8	35.5	95.9	10.5	50.9	—29	—20.2		
70	158.0	35	95.0	10	50.0	—30	—22.0		
69	156.2	34.5	94.1	9.5	49.1	—31	—23.8		
68	154.4	34	93.2	9	48.2	—32	—25.6		
67	152.6	33.5	92.3	8.5	47.3	—33	—27.4		
66	150.8	33	91.4	8	46.4	—34	—29.2		
65	149.0	32.5	90.5	7.5	45.5	—35	—31.0		
64	147.2	32	89.6	7	44.6	—36	—32.8		
63	145.4	31.5	88.7	6.5	43.7	—37	—34.6		
62	143.6	31	87.8	6	42.8	—38	—36.4		
61	141.8	30.5	86.9	5.5	41.9	—39	—38.2		
60	140.0	30	86.0	5	41.0	—40	—40.0		
59	138.2	29.5	85.1	4.5	40.1	—41	—41.8		
58	136.4	29	84.2	4	39.2	—42	—43.6		
57	134.6	28.5	83.3	3.5	38.3	—43	—45.4		
56	132.8	28	82.4	3	37.4	—44	—47.2		
55	131.0	27.5	81.5	2.5	36.5	—45	—49.0		
54	129.2	27	80.6	2	35.6	—46	—50.8		
53	127.4	26.5	79.7	1.5	34.7	—47	—52.6		
52	125.6	26	78.8	1	33.8	—48	—54.4		
51	123.8	25.5	77.9	0.5	32.9	—49	—56.2		

$x° \text{ Centesimal} = (32° + \tfrac{9}{5}x°) \text{ Fahrenheit.}$

TABLE VI.

TO COMPARE REAUMUR'S THERMOMETER WITH FAHRENHEIT'S.

Reaum.	Fahrenheit.	Reaum.	Fahrenheit.	Reaum.	Fahrenheit.	Reaum.	Fahrenheit.
o	o	o	o	o	o	o	o
80	212.0	40	122.00	20	77.00	0	+32.0
79	209.75	39.5	120.87	19.5	75.87	— 1	+29.75
78	207.5	39	119.75	19	74.75	— 2	+27.5
77	205.25	38.5	118.62	18.5	73.62	— 3	+25.25
76	203.0	38	117.50	18	72.50	— 4	+23.0
75	200.75	37.5	116.37	17.5	71.37	— 5	+20.75
74	198.5	37	115.25	17	70.25	— 6	+18.5
73	196.25	36.5	114.12	16.5	69.12	— 7	+16.25
72	194.0	36	113.00	16	68.00	— 8	+14.0
71	191.75	35.5	111.87	15.5	66.87	— 9	+11.75
70	189.5	35	110.75	15	65.75	—10	+ 9.5
69	187.25	34.5	109.62	14.5	64.62	—11	+ 7.25
68	185.0	34	108.50	14	63.50	—12	+ 5.0
67	182.75	33.5	107.37	13.5	62.37	—13	+ 2.75
66	180.5	33	106.25	13	61.25	—14	+ 0.5
65	178.25	32.5	105.12	12.5	60.12	—15	— 1.75
64	176.0	32	104.00	12	59.00	—16	— 4.0
63	173.75	31.5	102.87	11.5	57.87	—17	— 6.25
62	171.5	31	101.75	11	56.75	—18	— 8.5
61	169.25	30.5	100.62	10.5	55.62	—19	—10.75
60	167.0	30	99.50	10	54.50	—20	—13.0
59	164.75	29.5	98.37	9.5	53.37	—21	—15.25
58	162.5	29	97.25	9	52.25	—22	—17.5
57	160.25	28.5	96.12	8.5	51.12	—23	—19.75
56	158.0	28	95.00	8	50.00	—24	—22.0
55	155.75	27.5	93.87	7.5	48.87	—25	—24.25
54	153.5	27	92.75	7	47.75	—26	—26.5
•53	151.25	26.5	91.62	6.5	46.62	—27	—28.75
52	149.0	26	90.50	6	45.50	—28	—31.0
51	146.75	25.5	89.37	5.5	44.37	—29	—33.25
50	144.5	25	88.25	5	43.25	—30	—35.5
49	142.25	24.5	87.12	4.5	42.12	—31	—37.75
48	140.0	24	86.00	4	41.00	—32	—40.0
47	137.75	23.5	84.87	3.5	39.87	—33	—42.25
46	135.5	23	83.75	3	38.75	—34	—44.5
45	133.25	22.5	82.62	2.5	37.62	—35	—46.75
44	131.0	22	81.50	2	36.50	—36	—49.0
43	128.75	21.5	80.37	1.5	35.37	—37	—51.25
42	126.5	21	79.25	1	34.25	—38	—53.50
41	124.25	20.5	78.12	0.5	33.12	—39	—55.75
40	122.0	20	77.00		32.00	—40	—58.0

Proportional Parts.									
o	o	o	o	o	o	o	o	o	o
Reaumur....	0.1	0.2	0.3	0.4	0.5	0.6	0.7	0.8	0.9	1.0
Fahrenheit..	0.22	0.45	0.67	0.90	1.12	1.35	1.57	1.80	2.02	2.25

$$x^\circ \text{ Reaumur} = (32^\circ + \frac{9}{4}x^\circ) \text{ Fahrenheit.}$$

TABLE VII.—HEIGHT OF A COLUMN OF AIR, ETC. 257

TABLE VII.

HEIGHT OF A COLUMN OF AIR CORRESPONDING TO A TENTH OF AN INCH IN THE BAROMETER.

Barom.	40°	45°	50°	55°	60°	65°	70°	75°	80°	85°	90°
Inches.	Feet.	Feet.	Feet.	Feet.	Feet.	Feet.	Feet.	Feet.	Feet.	Feet.	Feet.
22.0	121.5	122.8	124.2	125.5	126.8	128.2	129.5	130.8	132.1	133.5	134.8
.2	120.4	121.7	123.1	124.4	125.7	127.0	128.3	129.6	130.9	132.2	133.6
.4	119.3	120.6	121.9	123.2	124.6	125.9	127.2	128.5	129.8	131.1	132.4
.6	118.2	119.5	120.8	122.1	123.4	124.7	126.0	127.3	128.6	129.9	131.2
.8	117.2	118.5	119.8	121.1	122.3	123.6	124.9	126.2	127.5	128.8	130.0
23.0	116.2	117.5	118.7	120.0	121.3	122.6	123.8	125.1	126.4	127.6	129.9
.2	115.2	116.5	117.7	119.0	120.2	121.5	122.7	124.0	125.3	126.5	127.8
.4	114.2	115.5	116.7	118.0	119.2	120.5	121.7	123.0	124.2	125.4	126.7
.6	113.2	114.4	115.7	116.9	118.1	119.4	120.6	121.8	123.1	124.3	125.5
.8	112.3	113.5	114.8	116.0	117.2	118.4	119.7	120.9	122.1	123.3	124.6
24.0	111.4	112.6	113.8	115.0	116.2	117.4	118.7	119.9	121.1	122.3	123.5
.2	110.5	111.7	112.9	114.1	115.3	116.5	117.7	118.9	120.1	121.3	122.5
.4	109.5	110.7	111.9	113.1	114.3	115.5	116.7	117.9	119.1	120.3	121.5
.6	108.6	109.8	111.0	112.2	113.4	114.6	115.8	116.9	118.1	119.3	120.5
.8	107.8	108.9	110.1	111.3	112.5	113.7	114.8	116.0	117.2	118.4	119.5
25.0	106.9	108.1	109.2	110.4	111.6	112.7	113.9	115.1	116.2	117.4	118.6
.2	106.0	107.2	108.4	109.5	110.7	111.8	113.0	114.1	115.3	116.5	117.6
.4	105.2	106.4	107.5	108.7	109.8	111.0	112.1	113.3	114.4	115.6	116.7
.6	104.4	105.5	106.7	107.8	108.9	110.1	111.2	112.4	113.5	114.6	115.8
.8	103.6	104.7	105.8	107.0	108.1	109.2	110.4	111.5	112.6	113.8	114.9
26.0	102.8	103.9	105.0	106.1	107.3	108.4	109.5	110.6	111.8	112.9	114.0
.2	102.0	103.1	104.2	105.3	106.5	107.6	108.7	109.8	110.9	112.0	113.1
.4	101.2	102.3	103.4	104.6	105.7	106.8	107.9	109.0	110.1	111.2	112.3
.6	100.5	101.6	102.7	103.8	104.9	106.0	107.1	108.2	109.3	110.4	111.4
.8	99.7	100.8	101.9	103.0	104.1	105.2	106.3	107.4	108.5	109.5	110.6
27.0	99.0	100.1	101.2	102.2	103.3	104.4	105.5	106.6	107.6	108.7	109.8
.2	98.3	99.3	100.4	101.5	102.6	103.6	104.7	105.8	106.8	107.9	109.0
.4	97.5	98.6	99.7	100.7	101.8	102.9	103.9	105.0	106.1	107.1	108.2
.6	96.8	97.9	98.9	100.0	101.1	102.1	103.2	104.2	105.3	106.3	107.4
.8	96.1	97.2	98.2	99.3	100.3	101.4	102.4	103.5	104.5	105.6	106.6
28.0	95.4	96.5	97.5	98.6	99.6	100.6	101.7	102.7	103.8	104.8	105.9
.2	94.8	95.8	96.8	97.9	98.9	99.9	101.0	102.0	103.0	104.1	105.1
.4	94.1	95.1	96.1	97.2	98.2	99.2	100.2	101.3	102.3	103.3	104.3
.6	93.4	94.4	95.5	96.5	97.5	98.5	99.5	100.6	101.6	102.6	103.6
.8	92.8	93.8	94.8	95.8	96.8	97.8	98.8	99.8	100.8	101.8	102.8
29.0	92.1	93.1	94.1	95.1	96.2	97.2	98.2	99.2	100.2	101.2	102.2
.2	91.5	92.5	93.5	94.5	95.5	96.5	97.5	98.5	99.5	100.5	101.5
.4	90.9	91.9	92.9	93.9	94.8	95.8	96.8	97.8	98.8	99.8	100.8
.6	90.3	91.3	92.2	93.2	94.2	95.2	96.2	97.2	98.2	99.1	100.1
.8	89.7	90.6	91.6	92.6	93.6	94.5	95.5	96.5	97.5	98.5	99.4
30.0	89.1	90.0	91.0	92.0	92.9	93.9	94.9	95.9	96.8	97.8	98.8
.2	88.5	89.4	90.4	91.4	92.3	93.3	94.3	95.2	96.2	97.2	98.1
.4	87.9	88.8	89.8	90.8	91.7	92.7	93.6	94.6	95.6	96.5	97.5
.6	87.3	88.2	89.2	90.2	91.1	92.1	93.0	94.0	95.0	95.9	96.8
.8	86.7	87.6	88.6	89.6	90.5	91.5	92.4	93.4	94.3	95.2	96.2

R

TABLE VIII.

FOR REDUCING BAROMETRIC OBSERVATIONS TO THE FREEZING POINT.

Temp.	27	27.5	28	28.5	29	29.5	30	30.5	31	Temp.
°										°
0	+.069	.071	.072	.073	.074	.076	.077	.078	.080	0
1	.067	.068	.069	.071	.072	.073	.074	.076	.077	1
2	.064	.066	.067	.068	.069	.070	.072	.073	.074	2
3	.062	.063	.064	.065	.067	.068	.069	.070	.071	3
4	.059	.061	.062	.063	.064	.065	.066	.067	.068	4
5	.057	.058	.059	.060	.061	.062	.063	.065	.066	5
6	+.055	.056	.057	.058	.059	.060	.061	.062	.063	6
7	.052	.053	.054	.055	.056	.057	.058	.059	.060	7
8	.050	.051	.052	.053	.054	.054	.055	.056	.057	8
9	.047	.048	.049	.050	.051	.052	.053	.054	.054	9
10	.045	.046	.047	.047	.048	.049	.050	.051	.052	10
11	+.042	.043	.044	.045	.046	.046	.047	.048	.049	11
12	.040	.041	.042	.042	.043	.044	.045	.045	.046	12
13	.038	.038	.039	.040	.040	.041	.042	.043	.043	13
14	.035	.036	.037	.037	.038	.038	.039	.040	.040	14
15	.033	.033	.034	.035	.035	.036	.036	.037	.038	15
16	+.030	.031	.032	.032	.033	.033	.034	.034	.035	16
17	.028	.028	.029	.030	.030	.031	.031	.032	.032	17
18	.025	.026	.026	.027	.027	.028	.028	.029	.029	18
19	.023	.024	.024	.024	.025	.025	.026	.026	.027	19
20	.021	.021	.021	.022	.022	.023	.023	.023	.024	20
21	+.018	.019	.019	.019	.020	.020	.020	.021	.021	21
22	.016	.016	.016	.017	.017	.017	.018	.018	.018	22
23	.013	.014	.014	.014	.014	.015	.015	.015	.015	23
24	.011	.011	.011	.012	.012	.012	.012	.012	.013	24
25	.009	.009	.009	.009	.009	.009	.009	.010	.010	25
26	+.006	.006	.006	.006	.007	.007	.007	.007	.007	26
27	+.004	.004	.004	.004	.004	.004	.004	.004	.004	27
28	+.001	.001	.001	.001	.001	.001	.001	.001	.001	28
29	—.001	.001	.001	.001	.001	.001	.001	.001	.001	29
30	—.004	.004	.004	.004	.004	.004	.004	.004	.004	30
31	—.006	.006	.006	.006	.007	.007	.007	.007	.007	31
32	.008	.009	.009	.009	.009	.009	.009	.010	.010	32
33	.011	.011	.011	.012	.012	.012	.012	.012	.012	33
34	.013	.014	.014	.014	.014	.015	.015	.015	.015	34
35	.016	.016	.016	.017	.017	.017	.018	.018	.018	35
36	—.018	.019	.019	.019	.020	.020	.020	.021	.021	36
37	.021	.021	.021	.022	.022	.022	.023	.023	.024	37
38	.023	.023	.024	.024	.025	.025	.026	.026	.026	38
39	.025	.026	.026	.027	.027	.028	.028	.029	.029	39
40	.028	.028	.029	.029	.030	.030	.031	.031	.032	40
41	—.030	.031	.031	.032	.033	.033	.034	.034	.035	41
42	.033	.033	.034	.034	.035	.036	.036	.037	.037	42
43	.035	.036	.036	.037	.038	.038	.039	.040	.040	43
44	.037	.038	.039	.040	.040	.041	.042	.042	.043	44
45	.040	.041	.041	.042	.043	.044	.044	.045	.046	45

TABLE VIII.

FOR REDUCING BAROMETRIC OBSERVATIONS TO THE FREEZING POINT.

Temp.	27	27.5	28	28.5	29	29.5	30	30.5	31	Temp.
45	—.040	.041	.041	.042	.043	.044	.044	.045	.046	45
46	.042	.043	.044	.045	.045	.046	.047	.048	.049	46
47	.045	.046	.046	.047	.048	.049	.050	.051	.051	47
48	.047	.048	.049	.050	.051	.052	.052	.053	.054	48
49	.050	.050	.051	.052	.053	.054	.055	.056	.057	49
50	.052	.053	.054	.055	.056	.057	.058	.059	.060	50
51	—.054	.055	.056	.057	.058	.059	.060	.061	.062	51
52	.057	.058	.059	.060	.061	.062	.063	.064	.065	52
53	.059	.060	.061	.063	.064	.065	.066	.067	.068	53
54	.062	.063	.064	.065	.066	.067	.068	.070	.071	54
55	.064	.065	.066	.068	.069	.070	.071	.072	.073	55
56	—.066	.068	.069	.070	.071	.073	.074	.075	.076	56
57	.069	.070	.071	.073	.074	.075	.076	.078	.079	57
58	.071	.073	.074	.075	.077	.078	.079	.081	.082	58
59	.074	.075	.076	.078	.079	.080	.082	.083	.085	59
60	.076	.077	.079	.080	.082	.083	.085	.086	.087	60
61	—.078	.080	.081	.083	.084	.086	.087	.089	.090	61
62	.081	.082	.084	.085	.087	.088	.090	.091	.093	62
63	.083	.085	.086	.088	.089	.091	.093	.094	.096	63
64	.086	.087	.089	.090	.092	.094	.095	.097	.098	64
65	.088	.090	.091	.093	.095	.096	.098	.100	.101	65
66	—.090	.092	.094	.096	.097	.099	.101	.102	.104	66
67	.093	.095	.096	.098	.100	.102	.103	.105	.107	67
68	.095	.097	.099	.101	.102	.104	.106	.108	.109	68
69	.098	.100	.101	.103	.105	.107	.109	.110	.112	69
70	.100	.102	.104	.106	.108	.109	.111	.113	.115	70
71	—.102	.104	.106	.108	.110	.112	.114	.116	.118	71
72	.105	.107	.109	.111	.113	.115	.117	.119	.120	72
73	.107	.109	.111	.113	.115	.117	.119	.121	.123	73
74	.110	.112	.114	.116	.118	.120	.122	.124	.126	74
75	.112	.114	.116	.118	.120	.122	.125	.127	.129	75
76	—.114	.117	.119	.121	.123	.125	.127	.129	.131	76
77	.117	.119	.121	.123	.126	.128	.130	.132	.134	77
78	.119	.122	.124	.126	.128	.130	.133	.135	.137	78
79	.122	.124	.126	.128	.131	.133	.135	.137	.140	79
80	.124	.126	.129	.131	.133	.136	.138	.140	.143	80
81	—.126	.129	.131	.134	.136	.138	.141	.143	.145	81
82	.129	.131	.134	.136	.138	.141	.143	.146	.148	82
83	.131	.134	.136	.139	.141	.143	.146	.148	.151	83
84	.134	.136	.139	.141	.144	.146	.149	.151	.154	84
85	.136	.139	.141	.144	.146	.149	.151	.154	.156	85
86	—.138	.141	.144	.146	.149	.151	.154	.156	.159	86
87	.141	.143	.146	.149	.151	.154	.157	.159	.162	87
88	.143	.146	.149	.151	.154	.157	.159	.162	.165	88
89	.146	.148	.151	.154	.156	.159	.162	.165	.167	89
90	.148	.151	.153	.156	.159	.162	.164	.167	.170	90

TABLE IX.

ALTITUDES WITH THE BAROMETER. — Part I.

Inches.	Feet.	Inches.	Feet.	Inches.	Feet.	Inches.	Feet.
11.0	1396.9	16.0	11186.3	21.0	18291.0	26.0	23871.0
.1	1633.3	.1	11349.1	.1	18415.1	.1	23971.3
.2	1867.6	.2	11510.9	.2	18538.7	.2	24071.2
.3	2099.9	.3	11671.7	.3	18661.6	,3	24170.7
.4	2330.1	.4	11831.5	..4	18784.0	.4	24269.8
.5	2558.3	.5	11990.3	.5	18905.8	.5	24368.6
.6	2784.5	.6	12148.2	.6	19027.0	.6	24467.0
.7	3008.7	.7	12305.1	.7	19147.7	.7	24565.1
.8	3231.1	.8	12461.0	.8	19267.8	.8	24662.7
.9	3451.6	.9	12616.1	.9	19387.4	.9	24760.0
12.0	3670.2	17.0	12770.2	22.0	19506.4	27.0	24857.0
.1	3887.0	.1	12923.5	.1	19624.9	.1	24953.6
.2	4102.0	.2	13075.8	.2	19742.9	.2	25049.8
.3	4315.3	.3	13227.3	.3	19860.3	.3	25145.7
.4	4526.9	.4	13377.9	.4	19977.2	.4	25241.2
.5	4736.7	.5	13527.6	.5	20093.6	.5	25336.4
.6	4944.9	.6	13676.5	.6	20209.4	.6	25431.2
.7	5151.4	.7	13824.5	.7	20324.8	.7	25525.7
.8	5356.4	.8	13971.7	.8	20439.6	.8	25619.9
.9	5559.7	.9	14118.0	.9	20554.0	.9	25713.7
13.0	5761.4	18.0	14263.6	23.0	20667.8	28.0	25807.1
.1	5961.6	.1	14408.3	.1	20781.1	.1	25900.3
.2	6160.2	.2	14552.3	.2	20894.0	.2	25993.1
.3	6357.5	.3	14695.4	.3	21006.4	.3	26085.6
.4	6553.2	.4	14837.8	.4	21118.3	.4	26177.7
.5	6747.5	.5	14979.4	.5	21229.7	.5	26269.6
.6	6940.3	.6	15120.3	.6	21340.6	.6	26361.1
.7	7131.7	.7	15260.3	.7	21451.1	.7	26452.3
.8	7321.7	.8	15399.7	.8	21561.1	.8	26543.2
.9	7510.3	.9	15538.3	.9	21670.6	.9	26633.7
14.0	7697.6	19.0	15676.2	24.0	21779.7	29.0	26724.0
.1	7883.6	.1	15813.3	.1	21888.4	.1	26813.9
.2	8068.2	.2	15949.8	.2	21996.6	.2	26903.5
.3	8251.5	.3	16085.5	.3	22104.3	.3	26992.8
.4	8433.6	.4	16220.5	.4	22211.6	.4	27081.9
.5	8614.4	.5	16354.8	.5	22318.4	.5	27170.6
.6	8794.0	.6	16488.5	.6	22424.8	.6	27259.0
.7	8972.3	.7	16621.4	.7	22530.8	.7	27347.1
.8	9149.5	.8	16753.7	.8	22636.4	.8	27434.9
.9	9325.5	.9	16885.3	.9	22741.5	.9	27522.5
15.0	9500.3	20.0	17016.3	25.0	22846.3	30.0	27609.7
.1	9673.8	.1	17146.6	.1	22950.6	.1	27696.6
.2	9846.2	.2	17276.3	.2	23054.4	.2	27783.3
.3	10017.5	.3	17405.3	.3	23157.9	.3	27869.7
.4	10187.7	.4	17533.7	.4	23261.0	.4	27955.7
.5	10356.8	.5	17661.4	.5	23363.6	.5	28041.5
.6	10524.8	.6	17788.6	.6	23465.9	.6	28127.1
.7	10691.8	.7	17915.1	.7	23567.7	.7	28212.3
.8	10857.7	.8	18041.0	.8	23669.2	.8	28297.3
.9	11022.5	.9	18166.3	.9	23770.3	.9	28382.0

TABLE IX.—ALTITUDES WITH THE BAROMETER. 261

TABLE IX.

ALTITUDES WITH THE BAROMETER — Part II.

T—T'.	Feet.	T—T'.	Feet.	T—T'.	Feet.	T—T'.	Feet.	T—T'.	Feet.
1°	2.3	17°	39.8	33°	77.3	49°	114.7	65°	152.2
2	4.7	18	42.1	34	79.6	50	117.0	66	154.5
3	7.0	19	44.5	35	81.9	51	119.4	67	156.8
4	9.4	20	46.8	36	84.3	52	121.7	68	159.2
5	11.7	21	49.2	37	86.6	53	124.1	69	161.5
6	14.0	22	51.5	38	89.0	54	126.4	70	163.9
7	16.4	23	53.8	39	91.3	55	128.7	71	166.2
8	18.7	24	56.2	40	93.6	56	131.1	72	168.6
9	21.1	25	58.5	41	96.0	57	133.4	73	170.9
10	23.4	26	60.9	42	98.3	58	135.8	74	173.3
11	25.8	27	63.2	43	100.7	59	138.1	75	175.6
12	28.1	28	65.5	44	103.0	60	140.4	76	177.9
13	30.4	29	67.9	45	105.3	61	142.8	77	180.3
14	32.8	30	70.2	46	107.7	62	145.1	78	182.6
15	35.1	31	72.6	47	110.0	63	147.5	79	185.0
16	37.5	32	74.9	48	112.4	64	149.8	80	187.3

Parts III., IV., and V.

Approximate Altitude	Part III. Positive from Lat. 0° to 45°. Negative from Lat. 45° to 90°. Latitude.						Part IV. Always positive.	Part V. Always positive. Height of Barometer at Lower Station.					
	0° / 90°	10° / 80°	20° / 70°	30° / 60°	40° / 50°	45°		18 in.	20 in.	22 in.	24 in.	26 in.	28 in.
Feet.	Feet.	Feet.	Feet.	Feet.	Feet.	Feet.	Feet.	Feet.	Feet.	Feet.	Feet.	Feet.	Feet.
1,000	2.6	2.5	2.0	1.3	0.5	0	2.5	1.3	1.0	0.8	0.6	0.4	0.2
2,000	5.3	5.0	4.1	2.6	0.9	0	5.2	2.5	2.0	1.5	1.1	0.7	0.3
3,000	7.9	7.5	6.1	4.0	1.4	0	7.9	3.8	3.0	2.3	1.7	1.1	0.5
4,000	10.6	10.0	8.1	5.3	1.8	0	10.8	5.1	4.0	3.1	2.2	1.4	0.7
5,000	13.2	12.4	10.1	6.6	2.3	0	13.7	6.4	5.0	3.8	2.8	1.8	0.8
6,000	15.9	14.9	12.2	7.9	2.8	0	16.7	7.6	6.0	4.6	3.3	2.1	1.0
7,000	18.5	17.4	14.2	9.3	3.2	0	19.9	8.9	7.1	5.4	3.9	2.5	1.2
8,000	21.2	19.9	16.2	10.6	3.7	0	23.1	10.2	8.1	6.2	4.4	2.8	1.3
9,000	23.8	22.4	18.3	11.9	4.1	0	26.4	11.4	9.1	6.9	5.0	3.2	1.5
10,000	26.5	24.9	20.3	13.2	4.6	0	29.8	12.7	10.1	7.7	5.5	3.5	1.7
11,000	29.1	27.4	22.3	14.6	5.1	0	33.3	14.0	11.1	8.5	6.1	3.9	1.8
12,000	31.8	29.9	24.4	15.9	5.5	0	36.9	15.3	12.1	9.2	6.6	4.2	2.0
13,000	34.4	32.4	26.4	17.2	6.0	0	40.6	16.5	13.1	10.0	7.2	4.6	2.2
14,000	37.1	34.9	28.4	18.5	6.4	0	44.4	17.8	14.1	10.8	7.7	4.9	2.3
15,000	39.7	37.3	30.4	19.9	6.9	0	48.3	19.1	15.1	11.5	8.3	5.3	2.5
16,000	42.4	39.8	32.5	21.2	7.4	0	52.3	20.3	16.1	12.3	8.8	5.6	2.7
17,000	45.0	42.3	34.5	22.5	7.8	0	56.4	21.6	17.1	13.1	9.4	6.0	2.8
18,000	47.7	44.8	36.5	23.8	8.3	0	60.5	22.9	18.1	13.8	9.9	6.3	3.0
19,000	50.3	47.3	38.6	25.2	8.7	0	64.8	24.1	19.2	14.6	10.5	6.7	3.2
20,000	53.0	49.8	40.6	26.5	9.2	0	69.2	25.4	20.2	15.4	11.0	7.0	3.3
21,000	55.6	52.3	42.6	27.8	9.7	0	73.6	26.7	21.2	16.1	11.6	7.4	3.5
22,000	58.3	54.8	44.7	29.1	10.1	0	78.2	28.0	22.2	16.9	12.1	7.7	3.7
23,000	60.9	57.3	46.7	30.5	10.6	0	82.9	29.2	23.2	17.7	12.7	8.1	3.8
24,000	63.6	59.8	48.7	31.8	11.0	0	87.6	30.5	24.2	18.5	13.2	8.4	4.0
25,000	66.2	62.2	50.7	33.1	11.5	0	92.5	31.8	25.2	19.2	13.8	8.8	4.1

TABLE X.

MEAN HEIGHT OF THE BAROMETER IN THE DIFFERENT MONTHS.

	George-town.	Havana.	Natchez.	St. Louis.	Philadel-phia.	Boston.	Toronto.	Port Bowen.	Renn. Harbor.
Jan...	29.942	30.129	29.779	29.602	29.961	29.976	29.618	29.716	29.773
Feb...	.965	29.928	.733	.586	.908	.957	.614	.886	.843
March	.957	.960	.723	.559	.942	.886	.622	30.107	.745
April.	.945	.905	.664	.490	.924	.897	.657	.068	.898
May..	.933	.851	.623	.465	.886	.907	.565	.051	.937
June .	.962	.948	.608	.478	.891	.878	.577	29.888	.714
July..	.971	.948	.646	.523	.915	.936	.589	.817	.736
Aug. .	.954	.816	.624	.542	.943	.965	.638	.683	.689
Sept. .	.943	.821	.749	.561	.971	'.994	.647	.689	.653
Oct...	.914	.851	.722	.588	.949	.964	.663	.784	.750
Nov...	.885	.971	.765	.588	.941	.955	.626	.899	.753
Dec...	.910	30.065	.816	.601	.959	.953	.643	.869	.748
Year	29.939	29.933	29.696	29.548	29.932	29.939	29.621	29.886	29.765

	Chris-tianborg.	Aden.	Cairo.	Constan-tinople.	Paris.	Green-wich.	St. Pe-tersburg.	Arch-angel.	Ham-merfest.
Jan...	29.829	29.823	30.016	30.009	29.877	29.760	30.022	29.728	29.605
Feb...	.840	.844	.016	29.982	.886	.804	.044	.743	.456
March	.874	.776	29.899	30.014	.777	.769	29.951	.698	.704
April.	.939	.701	.926	29.833	.732	.750	.969	.767	.816
May..	.971	.610	.852	.899	.749	.771	.958	.774	.888
June .	.958	.528	.702	.852	.814	.787	.915	.690	.818
July..	.920	.482	.682	.845	.785	.801	.853	.677	.816
Aug. .	.882	.512	.688	.860	.794	.789	.919	.672	.771
Sept. .	.862	.632	.792	.951	.788	.819	.969	.763	.740
Oct...	.849	.778	.910	30.033	.702	.696	.954	.713	.658
Nov...	.862	.876	.951	.104	.754	.754	.845	.680	.684
Dec...	.838	.890	.994	.014	.729	.821	.931	.675	.568
Year	29.885	29.704	29.856	29.943	29.782	29.777	29.944	29.711	29.710

	Singa-pore.	Madras.	Bombay.	Canton.	Benares.	Pekin.	Tiflis.	Nerts-chinsk.	Jakutsk.
Jan...	29.917	29.999	29.941	30.175	29.741	30.498	28.640	27.989	29.895
Feb...	.915	.970	.925	.099	.642	.429	.527	.960	.868
March	.884	.890	.872	.018	.575	.254	.534	.874	.749
April.	.886	.828	.799	29.849	.423	.091	.477	.744	.620
May..	.872	.709	.752	.761	.331	29.940	.468	.597	.472
June .	.858	.688	.649	.730	.179	.788	.421	.590	.366
July..	.868	.711	.656	.656	.161	.746	.392	.561	.383
Aug. .	.880	.747	.722	.659	.265	.860	.446	.650	.435
Sept. .	.886	.783	.791	.596	.370	30.079	.555	.758	.711
Oct...	.898	.863	.833	.911	.543	.268	.662	.858	.670
Nov...	.866	.943	.895	30.071	.649	.410	.689	.898	.828
Dec...	.884	.964	.944	.123	.744	.482	.628	.901	30.060
Year	29.884	29.842	29.815	29.895	29.468	30.154	28.538	27.782	29.679

TABLE XI.

MEAN HEIGHT OF THE BAROMETER FOR ALL HOURS OF THE DAY.

	Equator.	Cumana.	Calcutta.	Philadel-phia.	Toronto.	Padua.	Green-wich.	St. Pe-tersburg.	Van Renns'r.
Midnight	29.626	29.798	29.875	29.941	29.616	29.804	29.785	29.957	29.76t
1 A.M.	.615	.785	.868	.940	.615	.800	.778	.955	.763
2	.598	.772	.866	.938	.615	.798	.773	.954	.765
3	.592	.760	.863	.936	.615	.795	.770	.952	.766
4	.580	.751	.862	.938	.616	.794	.768	.952	.766
5	.593	.756	.861	.941	.621	.794	.768	.951	.766
6	.605	.771	.870	.951	.631	.796	.771	.953	.766
7	.626	.787	.888	.961	.639	.800	.777	.954	.766
8	.644	.803	.917	.967	.646	.804	.783	.957	.762
9	.652	.815	.926	.970	.648	.807	.789	.960	.762
10	.652	.816	.930	.968	.648	.810	.792	.961	.764
11	.638	.804	.926	.960	.641	.807	.791	.961	.764
Noon	.621	.787	.906	.943	.629	.805	.786	.959	.763
1 P.M.	.602	.764	.891	.927	.618	.798	.781	.956	.759
2	.589	.744	.859	.915	.608	.791	.776	.955	.759
3	.573	.731	.848	.908	.605	.786	.774	.954	.761
4	.568	.723	.839	.908	.603	.783	.772	.953	.763
5	.579	.731	.840	.910	.604	.782	.771	.953	.765
6	.595	.741	.843	.916	.608	.784	.774	.954	.767
7	.604	.756	.844	.925	.611	.789	.780	.955	.768
8	.617	.772	.865	.934	.616	.796	.785	.957	.769
9	.636	.788	.892	.941	.620	.801	.788	.957	.770
10	.640	.799	.895	.945	.620	.805	.790	.958	.771
11	.640	.810	.886	.950	.620	.805	.789	.958	.769
Mean	29.616	29.777	29.877	29.939	29.621	29.797	29.780	29.956	29.765

TABLE XII.

DEPRESSION OF MERCURY IN GLASS TUBES.

Diameter of Tube.	Ivory.	Young.	Laplace.	Poisson.	Caven-dish.	Pouillet.	Daniell, Boiled Tubes.	Diameter of Tube.
Inch.	Inch.	Inch.	Inch.	Inch.	Inch.	Inch.	Inch.	Inch.
0.05	0.2949	0.2964	0.	0.2796	0.	0.	0.	0.05
.10	.1404	.1424	.1394	.1367	.140	.1390	.070	.10
.15	.0865	.0880	.0854	.0830	.092	.0858	.044	.15
.20	.0583	.0589	.0580	.0559	.067	.0580	.029	.20
.25	.0409	.0404	.0412	.0394	.050	.0407	.020	.25
.30	.0293	.0280	.0296	.0281	.036	.0296	.014	.30
.35	.0212	.0196	.0216	.0204	.025	.0216	.010	.35
.40	.0154	.0139	.0159	.0149	.015	.0159	.007	.40
.45	.0112	.0100	.0117	.0109	.010	.0117	.005	.45
.50	.0082	.0074	.0087	.0080	.007	.0086	.003	.50
.60	.0043	.0045	.0046	.0041	.005	.0047	.002	.60
.70	.0023		.0024	.0020		.0025		.70
.80	.0012		.0013	.0010		.0013		.80

TABLE XIII.

TÓ COMPARE THE WEIGHT OF A CUBIC FOOT OF DRY AIR AND OF SAT-
URATED AIR.

Temp.	Dry Air.	Saturated.	Excess.	Temp.	Dry Air.	Saturated.	Excess.
Degrees.	Grains.	Grains.	Grains.	Degrees.	Grains.	Grains.	Grains.
0	603.21	602.77	0.44	45	548.16	546.06	2.10
1	601.87	601.40	0.47	46	547.05	544.88	2.17
2	600.52	600.03	0.49	47	545.97	543.75	2.22
3	599.20	598.69	0.51	48	544.85	542.55	2.30
4	597.87	597.34	0.53	49	543.75	541.36	2.39
5	596.55	596.01	0.54	50	542.65	540.21	2.44
6	595.24	594.69	0.55	51	541.55	539.04	2.51
7	593.94	593.36	0.58	52	540.48	537.87	2.61
8	592.63	592.04	0.59	53	539.41	536.71	2.70
9	591.33	590.72	0.61	54	538.33	535.55	2.78
10	590.04	589.40	0.64	55	537.27	534.39	2.88
11	588.75	588.07	0.68	56	536.19	533.22	2.97
12	587.48	586.78	0.70	57	535.12	532.06	3.06
13	586.21	585.49	0.72	58	534.07	530.92	3.15
14	584.93	584.18	0.75	59	533.03	529.77	3.26
15	583.67	582.89	0.78	60	531.97	528.62	3.35
16	582.41	581.61	0.80	61	530.93	527.48	3.45
17	581.15	580.33	0.82	62	529.88	526.32	3.56
18	579.91	579.06	0.85	63	528.84	525.17	3.67
19	578.67	577.79	0.88	64	527.81	524.03	3.78
20	577.44	576.54	0.90	65	526.78	522.90	3.88
21	576.21	575.27	0.94	66	525.76	521.75	4.01
22	574.98	574.01	0.97	67	524.75	520.61	4.14
23	573.76	572.76	1.00	68	523.72	519.46	4.26
24	572.55	571.50	1.05	69	522.70	518.29	4.41
25	571.33	570.26	1.07	70	521.70	517.17	4.53
26	570.13	569.01	1.12	71	520.70	516.02	4.68
27	568.92	567.77	1.15	72	519.69	514.87	4.82
28	567.73	566.53	1.20	73	518.70	513.75	4.95
29	566.54	565.31	1.23	74	517.70	512.61	5.09
30	565.35	564.08	1.27	75	516.71	511.46	5.25
31	564.17	562.86	1.31	76	515.73	510:32	5.41
32	563.00	561.64	1.36	77	514.74	509.18	5.56
33	561.84	560.42	1.42	78	513.77	508.04	5.73
34	560.67	559.20	1.47	79	512.80	506.91	5.89
35	559.51	558.01	1.50	80	511.82	505.74	6.08
36	558.35	556.79	1.56	81	510.87	504.61	6.26
37	557.21	555.61	1.60	82	509.89	503.45	6.44
38	556.05	554.40	1.65	83	508.93	502.32	6.61
39	554.91	553.20	1.71	84	507.97	501.16	6.81
40	553.77	552.00	1.77	85	507.03	500.05	6.98
41	552.65	550.80	1.84	86	506.07	498.87	7.20
42	551.52	549.63	1.89	87	505.11	497.71	7.40
43	550.39	548.44	1.95	88	504.19	496.58	7.61
44	549.27	547.26	2.01	89	503.25	495.44	7.81
45	548.16	546.06	2.10	90	502.32	494.28	8.04

TABLE XIV.—HEIGHT OF BAROMETER, ETC.   265

TABLE XIV.

HEIGHT OF BAROMETER CORRESPONDING TO TEMPERATURE OF BOILING WATER.

Temp.	Barom.	Temp.	Barom.	Temp.	Barom.	Temp.	Barom.	Temp.	Barom.
°	Inches.	°	Inches.	°	Inches.	°	Inches.	°	Inches.
188.0	18.197	193.0	20.253	198.0	22.501	203.0	24.952	208.0	27.622
.1	.236	.1	.296	.1	.548	.1	25.003	.1	.678
.2	.276	.2	.339	.2	.595	.2	.055	.2	.733
.3	.315	.3	.382	.3	.642	.3	.106	.3	.789
.4	.355	.4	.426	.4	.689	.4	.158	.4	.845
.5	.395	.5	.469	.5	.736	.5	.210	.5	.901
.6	.434	.6	.512	.6	.784	.6	.261	.6	.957
.7	.474	.7	.556	.7	.831	.7	.313	.7	28.013
.8	.514	.8	.599	.8	.879	.8	.365	.8	.069
.9	.554	.9	.643	.9	.926	.9	.417	.9	.126
189.0	.594	194.0	.687	199.0	.974	204.0	.469	209.0	.182
.1	.634	.1	.731	.1	23.022	.1	.521	.1	.239
.2	.674	.2	.775	.2	.070	.2	.573	.2	.295
.3	.714	.3	.819	.3	.118	.3	.626	.3	.352
.4	.755	.4	.863	.4	.166	.4	.678	.4	.409
.5	.795	.5	.907	.5	.214	.5	.730	.5	.466
.6	.835	.6	.951	.6	.262	.6	.783	.6	.523
.7	.876	.7	.996	.7	.311	.7	.836	.7	.580
.8	.917	.8	21.040	.8	.359	.8	.888	.8	.637
.9	.957	.9	.084	.9	.407	.9	.941	.9	.695
190.0	.998	195.0	.129	200.0	.456	205.0	.994	210.0	.752
.1	19.039	.1	.174	.1	.505	.1	26.047	.1	.810
.2	.080	.2	.218	.2	.553	.2	.100	.2	.867
.3	.121	.3	.263	.3	.602	.3	.153	.3	.925
.4	.162	.4	.308	.4	.651	.4	.206	.4	.983
.5	.203	.5	.353	.5	.700	.5	.259	.5	29.041
.6	.244	.6	.398	.6	.749	.6	.313	.6	.099
.7	.285	.7	.443	.7	.798	.7	.366	.7	.157
.8	.326	.8	.488	.8	.847	.8	.420	.8	.215
.9	.368	.9	.533	.9	.897	.9	.473	.9	.274
191.0	.409	196.0	.578	201.0	.946	206.0	.527	211.0	.332
.1	.450	.1	.623	.1	.996	.1	.581	.1	.391
.2	.492	.2	.669	.2	24.045	.2	.635	.2	.449
.3	.534	.3	.714	.3	.095	.3	.689	.3	.508
.4	.575	.4	.760	.4	.145	.4	.743	.4	.567
.5	.617	.5	.806	.5	.195	.5	.797	.5	.626
.6	.659	.6	.851	.6	.245	.6	.852	.6	.685
.7	.701	.7	.897	.7	.295	.7	.906	.7	.744
.8	.743	.8	.943	.8	.345	.8	.961	.8	.803
.9	.785	.9	.989	.9	.395	.9	27.015	.9	.863
192.0	.827	197.0	22.033	202.0	.445	207.0	.070	212.0	.922
.1	.869	.1	.081	.1	.495	.1	.125	.1	.982
.2	.912	.2	.128	.2	.546	.2	.180	.2	30.041
.3	.954	.3	.174	.3	.596	.3	.235	.3	.101
.4	.996	.4	.221	.4	.647	.4	.290	.4	.161
.5	20.039	.5	.267	.5	.697	.5	.345	.5	.221
.6	.082	.6	.314	.6	.748	.6	.400	.6	.281
.7	.124	.7	.361	.7	.799	.7	.456	.7	.341
.8	.167	.8	.407	.8	.850	.8	.511	.8	.401
.9	.210	.9	.454	.9	.901	.9	.566	.9	.462

TABLE XV.

DIURNAL VARIATION OF TEMPERATURE AT NEW HAVEN, CONNECTICUT.

Hours.	Jan.	Feb.	Mar.	Apr.	May.	June.	July.	Aug.	Sept.	Oct.	Nov.	Dec.	Year.
	o	o	o	o	o	o	o	o	o	o	o	o	o
Midn'ht	2.3	2.9	3.7	4.6	5.4	5.8	5.2	4.7	4.8	4.1	2.6	2.2	4.0
1	2.6	3.3	4.3	5.4	6.3	6.9	6.2	5.6	5.6	4.8	3.2	2.5	4.7
2	3.0	3.8	4.8	6.1	7.2	8.0	7.0	6.3	6.3	5.5	3.7	2.8	5.4
3	3.3	4.3	5.4	6.7	8.0	8.7	7.5	6.8	6.8	6.0	4.1	3.2	5.9
4	3.7	4.8	5.8	7.3	8.5	8.9	7.7	7.2	7.2	6.5	4.5	3.5	6.3
5	4.1	5.2	6.2	7.5	8.4	8.2	7.4	7.1	7.3	6.8	4.8	3.8	6.4
6	4.3	5.3	6.1	7.1	6.6	6.1	6.1	6.4	6.8	6.6	4.8	4.0	5.9
7	4.4	5.1	4.9	5.3	3.6	3.2	3.7	4.1	4.7	5.3	4.5	4.0	4.4
8	3.8	3.7	2.3	2.0	0.5	0.0	0.9	1.3	1.7	2.3	3.0	3.2	2.1
9	1.3	0.5	-0.5	-1.1	-2.1	-2.7	-1.6	-1.2	-1.2	-0.6	0.5	1.0	-0.6
10	-1.6	-2.5	-3.2	-3.9	-4.2	-4.7	-3.8	-3.4	-3.8	-3.5	-2.2	-1.6	-3.2
11	-3.6	-4.2	-4.9	-5.5	-5.8	-6.1	-5.6	-5.3	-5.6	-5.6	-4.2	-3.5	-5.0
Noon	-5.2	-5.6	-6.1	-6.8	-7.0	-7.1	-6.7	-6.5	-6.9	-7.0	-5.6	-5.0	-6.3
1	-6.1	-6.6	-7.0	-7.7	-7.9	-7.9	-7.5	-7.3	-7.7	-7.7	-6.4	-5.8	-7.1
2	-6.3	-6.9	-7.5	-8.3	-8.5	-8.3	-7.8	-7.7	-8.0	-8.1	-6.6	-6.1	-7.5
3	-5.9	-6.8	-7.3	-8.3	-8.5	-8.2	-7.7	-7.6	-7.9	-7.7	-6.2	-5.5	-7.3
4	-4.7	-5.8	-6.6	-7.8	-8.0	-7.6	-7.2	-7.1	-7.1	-6.6	-4.6	-4.0	-6.4
5	-2.8	-3.8	-4.7	-6.6	-6.8	-6.5	-6.1	-5.9	-5.8	-4.5	-2.9	-2.1	-4.9
6	-1.4	-2.0	-2.5	-4.0	-4.7	-4.3	-4.2	-3.9	-4.0	-2.8	-1.6	-1.0	-3.0
7	-0.3	-0.6	-0.9	-1.5	-1.6	-2.2	-2.0	-1.9	-1.9	-1.2	-0.5	-0.2	-1.2
8	0.5	0.4	0.6	0.6	0.6	0.1	-0.1	0.0	0.1	0.2	0.4	0.5	0.3
9	1.1	1.2	1.7	2.0	2.2	1.8	1.6	1.6	1.7	1.5	1.1	1.0	1.5
10	1.5	1.8	2.4	3.0	3.5	3.3	2.9	2.8	2.8	2.4	1.6	1.5	2.5
11	1.9	2.4	3.0	3.8	4.4	4.6	4.1	3.9	3.9	3.3	2.1	1.8	3.3
Da. ext.	-1.0	-0.8	-0.6	-0.4	0.0	0.3	-0.1	-0.3	-0.3	-0.6	-0.9	-1.1	-0.6
7.1	-0.8	-0.7	-1.1	-1.2	-2.1	-2.4	-1.9	-1.6	-1.5	-1.2	-0.9	-0.9	-1.4
7.2	-1.0	-0.9	-1.3	-1.5	-2.4	-2.6	-2.1	-1.8	-1.6	-1.4	-1.0	-1.1	-1.6
8.1	-1.1	-1.4	-2.4	-2.8	-3.7	-4.0	-3.3	-3.0	-3.0	-2.7	-1.7	-1.3	-2.5
8.2	-1.3	-1.6	-2.6	-3.1	-4.0	-4.2	-3.5	-3.2	-3.2	-2.9	-1.8	-1.4	-2.7
9.8	0.9	0.5	0.0	-0.2	-0.8	-1.3	-0.9	-0.6	-0.6	-0.2	0.4	0.7	-0.2
6.6	1.5	1.6	1.8	1.5	0.9	0.9	1.0	1.2	1.4	1.9	1.6	1.5	1.4
7.7	2.0	2.2	2.0	1.9	1.0	0.5	0.8	1.1	1.4	2.0	2.0	1.9	1.6
8.8	2.1	2.1	1.4	1.3	0.6	0.0	0.4	0.7	0.9	1.2	1.7	1.8	1.2
9.9	1.2	0.9	0.6	0.4	0.0	-0.4	0.0	0.2	0.2	0.4	0.8	1.0	0.4
10.10	0.0	-0.3	-0.4	-0.4	-0.4	-0.7	-0.4	-0.3	-0.5	-0.5	-0.3	-0.1	-0.4
6.2.6	-1.1	-1.2	-1.3	-1.7	-2.2	-2.2	-1.9	-1.8	-1.7	-1.4	-1.1	-1.0	-1.6
6.2.8	-0.5	-0.4	-0.3	-0.2	-0.4	-0.7	-0.6	-0.4	-0.4	-0.4	-0.5	-0.5	-0.4
6.2.9	-0.3	-0.1	0.1	0.3	0.1	-0.1	0.0	0.1	0.2	0.0	-0.2	-0.4	0.0
6.2.10	-0.1	0.1	0.3	0.6	0.5	0.4	0.4	0.5	0.5	0.3	-0.1	-0.2	0.3
7.2.9	-0.3	-0.2	-0.3	-0.4	-0.9	-1.1	-0.8	-0.6	-0.5	-0.4	-0.3	-0.4	-0.5
7.2.9.9	0.1	0.1	0.2	0.2	-0.1	-0.4	-0.2	-0.1	0.0	0.0	0.0	0.0	0.0

TABLE XVI.—DIURNAL VARIATION OF TEMPERATURE. 267

TABLE XVI.

DIURNAL VARIATION OF TEMPERATURE AT GREENWICH, ENGLAND.

Hours.	Jan.	Feb.	Mar.	Apr.	May.	June.	July.	Aug.	Sept.	Oct.	Nov.	Dec.	Year.
	o	o	o	o	o	o	o	o	o	o	o	o	o
Midn'ht	1.0	1.6	2.9	4.8	5.4	6.2	5.0	5.1	4.0	2.9	1.7	0.9	3.5
1	0.9	1.8	3.0	5.2	6.0	7.1	5.5	5.5	4.5	3.0	1.8	1.0	3.8
2	1.2	2.0	3.3	5.7	6.4	8.0	6.0	6.0	5.5	3.4	2.0	1.2	4.2
3	1.3	2.1	3.6	6.2	6.7	8.7	6.4	6.3	6.4	3.6	2.0	1.3	4.5
4	1.6	2.3	3.9	6.6	6.7	9.3	6.6	6.5	6.6	3.8	2.1	1.4	4.8
5	1.8	2.2	4.0	6.7	6.3	8.8	6.2	6.5	6.2	3.8	2.0	1.4	4.7
6	1.9	2.3	3.9	6.0	4.8	6.4	4.5	5.5	5.3	3.5	1.9	1.4	3.9
7	1.9	2.1	3.6	4.3	2.6	3.0	2.5	3.3	4.0	2.8	1.7	1.5	2.8
8	1.5	1.6	2.5	2.0	0.5	0.0	0.0	0.9	2.1	1.6	1.0	1.3	1.2
9	1.0	0.7	0.2	-0.9	-2.0	-2.5	-2.0	-1.6	-0.4	0.0	0.4	0.9	-0.5
10	0.2	-0.5	-1.9	-3.2	-4.0	-4.5	-4.0	-3.5	-3.0	-2.0	-0.6	0.0	-2.2
11	-1.3	-2.1	-3.5	-5.3	-5.5	-5.8	-5.4	-5.4	-5.0	-3.8	-2.0	-1.3	-3.9
Noon	-2.3	-3.2	-5.0	-6.8	-6.7	-7.3	-6.4	-6.5	-6.4	-5.1	-3.1	-2.1	-5.1
1	-2.9	-3.9	-5.8	-7.9	-7.5	-8.1	-6.7	-7.5	-7.1	-5.5	-3.5	-2.4	-5.7
2	-3.0	-3.9	-5.8	-8.2	-7.7	-8.6	-6.7	-7.7	-7.1	-4.9	-3.6	-2.3	-5.8
3	-2.5	-3.6	-5.5	-7.7	-7.3	-8.4	-6.5	-7.0	-6.6	-3.7	-3.0	-1.9	-5.3
4	-1.9	-2.8	-4.5	-6.7	-6.1	-7.4	-5.8	-5.5	-5.5	-2.8	-2.1	-1.3	-4.4
5	-1.1	-1.6	-3.3	-5.4	-4.8	-6.1	-4.9	-3.6	-4.2	-1.7	-1.2	-0.8	-3.2
6	-0.6	-0.6	-1.8	-3.5	-3.0	-4.5	-3.5	-2.0	-2.5	-0.8	-0.4	-0.4	-2.0
7	-0.3	0.3	-0.4	-1.1	-1.0	-2.4	-1.5	-0.5	-0.6	0.0	0.1	-0.1	-0.6
8	0.1	0.6	0.9	0.7	0.9	0.0	0.3	1.0	1.0	0.7	0.6	0.2	0.6
9	0.4	1.0	1.7	2.0	2.3	1.8	1.9	2.4	1.8	1.3	1.0	0.4	1.5
10	0.6	1.3	2.3	3.2	3.5	3.6	3.3	3.3	2.7	1.9	1.3	0.5	2.3
11	0.7	1.5	2.6	4.1	4.5	5.0	4.2	4.3	3.4	2.4	1.5	0.8	2.9
Da. ext.	-0.5	-0.8	-0.9	-0.7	-0.5	0.3	-0.1	-0.6	-0.3	-0.8	-0.7	-0.4	-0.5
7.1	-0.5	-0.9	-1.1	-1.8	-2.4	-2.6	-2.1	-2.1	-1.5	-1.4	-0.9	-0.4	-1.5
7.2	-0.5	-0.9	-1.1	-1.9	-2.5	-2.8	-2.1	-2.2	-1.5	-1.0	-0.9	-0.4	-1.5
8.1	-0.7	-1.1	-1.6	-2.9	-3.5	-4.0	-3.4	-3.3	-2.5	-1.7	-1.3	-0.5	-2.2
8.2	-0.7	-1.1	-1.6	-3.1	-3.6	-4.3	-3.3	-3.4	-2.5	-1.7	-1.3	-0.5	-2.3
9.8	0.5	0.6	0.5	-0.1	-0.5	-1.2	-0.8	-0.3	0.3	0.3	0.5	0.5	0.0
6.6	0.6	0.9	1.0	1.2	0.9	0.9	0.5	1.7	1.4	1.3	0.8	0.5	0.9
7.7	0.8	1.2	1.6	1.6	0.8	0.3	0.5	1.4	1.7	1.4	0.9	0.7	1.1
8.8	0.8	1.1	1.7	1.3	0.7	0.0	0.1	0.9	1.5	1.1	0.8	0.8	0.9
9.9	0.7	0.8	0.9	0.5	0.1	-0.3	0.0	0.4	0.7	0.6	0.7	0.6	0.5
10.10	0.4	0.4	0.2	0.0	-0.2	-0.4	-0.4	-0.1	-0.1	0.0	0.4	0.2	0.0
6.2.6	-0.6	-0.7	-1.2	-1.9	-1.9	-2.2	-1.9	-1.4	-1.4	-0.7	-0.7	-0.4	-1.3
6.2.8	-0.3	-0.3	-0.3	-0.5	-0.7	-0.7	-0.6	-0.4	-0.3	-0.2	-0.4	-0.2	-0.4
6.2.9	-0.2	-0.1	-0.1	-0.1	-0.2	-0.1	-0.1	0.0	0.0	0.0	-0.2	-0.2	-0.1
6.2.10	-0.2	-0.1	0.1	0.3	0.2	0.5	0.4	0.3	0.3	0.2	-0.1	-0.1	0.1
7.2.9	-0.2	-0.3	-0.2	-0.6	-0.9	-1.2	-0.8	-0.7	-0.4	-0.2	-0.3	-0.1	-0.5
7.2.9.9	-0.1	0.1	0.3	0.0	-0.1	-0.5	-0.1	-0.1	0.1	0.1	0.0	0.0	0.0

TABLE XVII.

MEAN TEMPERATURES FOR EACH MONTH, SEASON, AND THE YEAR.

Place.	Lat.	Long.	Alt.	Jan.	Feb.	March.	April.
	° ′	° ′	Feet.	°	°	°	°
Paramaribo, Dutch Guiana ..	5 44	55 13		78.2	78.0	78.9	79.2
St. Vincent, West Indies.....	13 10	60 31		80.0	79.3	79.7	81.3
Kingston, Jamaica.........	17 58	76 50	50	75.7	76.0	75.9	78.1
Vera Cruz, Mexico...........	19 12	96 9	50	70.0	71.6	73.4	77.2
Mexico City..............	19 25	99 6	6990	52.5	56.4	61.1	63.0
Havana, Cuba.............	23 9	82 23	50	71.4	74.0	74.1	76.6
Key West, Florida.........	24 32	81 47	10	68.1	69.4	72.7	75.3
Galveston, Texas..........	29 18	94 47		54.2	60.2	69.2	71.6
St. Augustine, Florida......	29 48	81 35	20	57.0	59.9	63.3	68.8
New Orleans, Louisiana.....	29 57	90 0	10	54.8	56.4	62.9	69.0
Bermuda, Atlantic Ocean....	32 20	64 50		56.8	58.8	59.4	62.8
San Diego, California.......	32 42	117 13	150	51.9	53.3	56.0	61.2
Charleston, South Carolina...	32 46	79 56	20	50.3	52.4	58.7	65.4
Santa Fé, New Mexico.....	35 41	106 1	6846	31.4	33.2	40.7	51.3
Richmond, Virginia........	37 32	77 27	120	33.7	39.8	47.1	54.7
San Francisco, California....	37 48	122 26	100	49.8	52.3	53.7	57.0
St. Louis, Missouri.........	38 37	90 15	450	32.9	35.0	44.4	58.3
Washington, D. C..........	38 53	77 0	80	34.1	36.7	45.3	55.7
Cincinnati, Ohio	39 6	84 30	543	33.1	34.1	43.5	54.1
Baltimore, Maryland	39 18	76 37	36	32.8	34.2	42.3	52.7
Philadelphia, Pennsylvania ..	39 58	75 10	40	31.8	32.3	41.0	51.8
New York City, New York ..	40 43	74 0	23	30.2	30.4	38.3	48.6
Salt Lake City, Utah.......	40 46	112 6	4351	27.1	34.0	39.7	50.2
New Haven, Connecticut....	41 18	72 55	50	26.5	28.1	36.1	46.8
Cleveland, Ohio...........	41 31	81 51	660	26.2	29.0	35.6	47.7
Chicago, Illinois	41 54	87 38	591	23.6	24.7	32.3	46.1
Fort Laramie, Dacotah.....	42 12	104 48	4519	31.0	32.6	36.8	47.6
Detroit, Michigan	42 20	83 2	580	27.0	26.6	35.4	46.3
Boston, Massachusetts......	42 21	71 3	50	27.8	27.9	36.2	46.4
Buffalo, New York.........	42 53	78 55	600	23.4	21.1	35.5	40.7
Toronto, Canada ....·......	43 40	79 23	341	24.3	23.1	30.4	41.3
Halifax, Nova Scotia.......	44 39	63 37	20	22.6	23.7	30.9	38.9
Fort Snelling, Wisconsin....	44 53	93 8	820	13.7	17.6	31.4	46.3
Astoria, Oregon...........	46 11	123 48	50	43.0	43.6	45.7	52.8
Fort Brady, Michigan......	46 39	84 43	600	17.2	16.2	25.1	38.3
Quebec, Canada...........	46 49	71 12	300	10.4	13.8	26.4	39.0
St. Johns, Newfoundland....	47 33	52 28	140	23.3	20.9	24.2	33.4
Cumberland House	53 57	102 17	900	− 7.0	− 4.6	15.2	31.0
Sitka, Aliashka...........	57 3	135 18	50	30.0	30.7	34.1	39.9
Nain, Labrador...........	57 10	61 50	50	− 2.9	− 0.7	7.6	22.7
Godhaab, Greenland	64 10	52 24		12.4	12.6	15.6	22.0
Fort Franklin, Brit. America .	65 12	123 13	500	−22.3	−16.7	− 5.4	12.4
Boothia Felix, Arctic Regions	69 59	92 1		−28.7	−32.0	−28.7	− 2.6
Melville Island, Arc. Regions.	74 47	110 48		−31.3	−32.4	−18.2	− 8.2
Van Rensselaer Harbor, A. R.	78 37	70 53	5	−28.2	−26.4	−34.9	−10.3

TABLE XVII.

MEAN TEMPERATURES FOR EACH MONTH, SEASON, AND THE YEAR.

May.	June.	July.	Aug.	Sept.	Oct.	Nov.	Dec.	Spring.	Summer.	Autumn.	Winter.	Year.
°	°	°	°	°	°	°	°	°	°	°	°	°
79.9	79.5	80.0	82.0	83.4	83.3	81.5	79.7	79.3	80.5	82.7	78.6	80.3
82.2	82.2	82.2	82.8	83.3	82.7	82.1	80.4	81.1	82.4	82.7	79.9	81.5
80.3	80.6	81.7	81.0	80.7	79.8	78.7	76.7	76.2	78.1	81.1	79.7	78.8
80.4	81.9	81.5	82.4	81.0	78.4	75.4	71.1	77.0	81.9	78.3	70.9	77.0
66.1	65.4	65.4	64.9	64.3	60.2	55.8	52.0	63.4	65.2	60.1	53.6	60.6
78.0	81.0	81.5	81.6	80.4	78.8	75.1	73.5	76.2	81.3	78.1	73.0	77.2
79.0	81.4	82.7	82.8	81.6	77.7	74.7	70.7	75.7	82.3	78.0	69.7	76.4
81.1	83.5	85.6	85.9	82.7	68.6	60.1	57.5	74.0	85.0	70.5	57.3	71.7
73.5	79.3	80.9	80.5	78.6	71.9	64.1	57.2	68.5	80.3	71.5	58.1	69.6
74.8	79.9	81.6	81.2	78.5	69.8	60.2	56.0	68.9	81.0	69.3	55.7	68.7
69.1	73.2	75.7	76.6	76.8	73.0	65.8	60.6	63.7	75.2	71.9	58.8	67.4
62.7	67.4	72.7	73.7	70.9	65.5	56.9	51.7	60.0	71.2	64.4	52.3	62.0
73.4	79.0	81.7	80.9	76.9	67.9	59.5	52.5	65.8	80.6	68.1	51.7	66.6
57.1	68.8	72.6	70.0	61.9	51.3	38.6	30.2	49.7	70.4	50.6	31.6	50.6
65.4	73.8	77.6	74.8	67.1	57.5	44.2	38.1	55.7	75.4	56.3	37.2	56.2
56.5	57.7	58.8	59.0	59.9	59.8	55.6	51.3	55.7	58.7	58.4	51.2	56.0
66.4	74.0	78.5	76.5	68.7	55.4	40.9	33.6	56.4	76.3	55.0	33.8	55.4
66.3	74.4	78.3	76.3	67.7	56.7	44.8	37.3	55.8	76.3	56.4	36.1	56.1
63.6	71.4	76.5	74.2	66.0	53.2	42.5	33.8	53.7	74.0	53.9	33.7	53.8
63.1	71.6	76.7	74.7	67.8	55.7	45.1	35.6	52.7	74.3	56.2	34.2	54.3
62.5	71.5	76.0	73.2	63.8	54.5	44.0	34.5	51.8	73.6	54.1	32.9	53.1
59.3	68.3	74.8	73.2	65.8	54.5	43.3	33.5	48.7	72.1	54.5	31.4	51.7
63.0	71.3	76.9	75.0	67.1	55.6	41.7	31.3	51.0	74.4	54.8	30.8	52.7
57.3	67.0	71.7	70.3	62.5	51.1	40.3	30.4	46.7	70.0	51.3	28.4	49.0
56.6	66.3	71.9	68.8	62.4	49.3	37.9	29.6	46.6	69.0	49.9	28.3	48.5
56.3	62.7	70.8	68.5	60.1	48.5	37.9	29.3	44.9	67.3	48.8	25.9	46.7
56.1	67.3	74.7	73.8	64.2	50.9	35.8	28.0	46.8	71.9	50.3	31.1	50.1
56.0	65.6	69.7	67.5	60.0	47.7	38.2	26.9	45.9	67.6	48.7	26.8	47.2
56.5	66.2	71.6	69.4	62.2	51.5	41.0	31.1	46.3	69.1	51.6	28.9	48.9
55.3	67.4	71.5	70.0	59.9	48.7	37.2	22.8	43.8	69.6	48.6	22.4	46.1
51.5	61.4	66.8	66.3	58.1	45.2	36.6	26.2	41.1	64.8	46.6	24.5	44.3
48.0	56.0	62.0	64.4	58.4	48.0	38.5	27.7	39.3	60.8	48.3	24.7	43.2
59.0	68.4	73.4	70.1	58.9	47.1	31.7	16.9	45.6	70.6	45.9	16.1	44.6
55.0	59.5	61.6	63.0	58.7	55.4	46.4	40.7	51.1	61.6	53.7	42.4	52.2
49.3	58.4	64.7	62.9	54.6	43.5	32.5	21.5	37.6	62.0	43.5	18.3	40.4
53.2	64.5	69.0	68.1	56.8	43.9	32.9	15.0	39.5	67.2	44.6	13.0	40.6
39.3	48.0	56.2	57.9	53.0	44.5	34.0	25.3	32.3	54.0	43.8	23.2	38.3
51.3	58.8	61.8	59.5	45.8	35.0	17.2	5.6	32.5	60.0	32.7	-2.0	30.8
46.0	52.5	55.1	55.1	50.0	44.1	37.7	35.9	40.0	54.2	43.9	32.2	42.6
32.8	41.8	48.2	51.1	42.2	32.2	22.3	3.4	21.7	47.0	32.2	-0.4	25.1
32.2	39.1	41.9	40.7	35.6	22.9	17.5		23.3	40.6	29.9	14.1	26.8
35.2	48.0	52.1	50.6	41.0	22.5	-0.1	-10.9	14.0	50.2	21.1	-16.7	17.2
15.6	34.2	41.3	38.7	25.4	9.1	-5.4	-22.4	-5.2	38.0	9.7	-27.7	3.7
16.8	36.2	42.4	32.6	22.5	-2.8	-21.1	-21.6	-3.2	37.1	-0.5	-28.4	1.2
13.4	30.1	38.2	31.8	13.4	-3.6	-21.9	-31.1	-10.6	33.4	-4.0	-28.6	-2.5

TABLE XVIII.—PLACES WHOSE MEAN TEMPERATURE IS ABOVE 80° FAH.

Place.	Latitude.	Longitude.	Tempera-ture.	No. of Years.
	° ′	° ′	°	
Niger, Africa...............	5 9	— 6	85.27	1
Maracaibo, South America....	10 43	71 52	84.75	1
Kouka, Central Africa.......	13 10	— 14 30	83.63	2
Coburg Peninsula, Australia...	—11 5	—132 15	82.79	1
Pondichery, India..........	11 56	— 79 52	82.58	2
Calcutta, India............	22 35	— 88 20	82.41	4
Madras, India..............	13 4	— 80 19	81.94	20
Samarang, Java............	— 6 50	—110 30	81.87	1
Nagpoor, India............	21 8	— 79 11	81.59	3
Rio Berbice, British Guiana...	6 29	57 24	81.56	1
St. Vincent, West Indies......	13 10	60 31	81.52	6
Porto Rico, West Indies......	18 29	66 13	81.38	5
Guinea, Africa.............	5 30	0 0	81.38	1
St. Domingo, West Indies.....	18 29	70 0	81.29	1
St. Christopher, West Indies...	17 44	64 49	81.27	1
Bombay, India.............	18 56	— 72 54	81.27	1
St. Thomas, West Indies......	18 21	64 56	81.23	1
Anjarakandy, India.........	11 40	— 75 40	81.07	10
Cobbe, Africa..............	14 11	— 28 8	80.96	2
Colombo, Ceylon...........	6 57	— 80 0	80.75	1
Trincomalee, Ceylon.........	8 34	— 81 22	80.75	2
Demerara, British Guiana.....	6 45	58 2	80.71	1
Para, Brazil...............	— 1 28	48 29	80.70	4
Singapore, Malacca..........	1 17	—103 50	80.68	6
Upper Park Camp, Jamaica...	17 58	76 50	80.63	1
Fort Dundas, Australia.......	—11 25	—132 25	80.63	1
Christiansborg, Africa.......	5 24	— 0 16	80.42	4
Paramaribo, Dutch Guiana....	5 44	55 13	80.30	2
Benares, India.............	25 18	— 82 56	80.26	3
Kingstown, West Indies......	13 8	60 37	80.25	1
Cawnpore, India	26 29	— 80 22	80.21	1
Upper Egypt...............	26 0	— 33 40	80.10	1

TABLE XIX.—PLACES WHOSE MEAN TEMPERATURE IS BELOW 18° FAH.

Place.	Latitude.	Longitude.	Tempera-ture.	No. of Years.
	° ′	° ′	°	
Van Rensselaer Harbor.......	78 37	70 53	—2.46	2
Melville Island.............	74 47	110 48	+1.24	1
Ustjansk, Siberia	70 55	—138 24	2.75	2
Port Bowen, Arctic Regions...	73 14	88 56	3.53	1
Boothia Felix, Arctic Regions..	69 59	92 1	3.70	1
Igloolik, North America......	69 21	81 53	5.55	1
Fort Hope, North America....	62 32	86 56	6.10	
Winter Island	66 11	83 11	8.82	1
Nishne Kolymsk, Siberia	68 32	—160 56	9.50	2
Jakutsk, Siberia............	62 2	—129 44	11.53	17
Fort Enterprise, North America	64 20	113 6	13.90	3
Karische Pforte, Nova Zembla	70 36	— 57 47	14.90	1
Yucon, Russian America.....	66 0	147 0	16.80	1
Matoshkin Schar, Nova Zembla	73 19	— 57 20	16.93	1
Fort Franklin, Great Bear Lake	65 12	123 13	17.18	2
Fort Churchill, Hudson Bay...	59 2	93 10	17.45	2

TABLE XX.—PLACES HAVING A SMALL MONTHLY RANGE OF TEMPERATURE.

Place.	Latitude.	Longitude.	Hottest Month.	Coldest Month.	Differ- ence.	No. of Years.
	o ′	o ′	o	o	o	
Commewine, South America..	5 38	54 42	79.2	77.0	2.2	2
Buitenzorg, Java..........	— 6 37	—106 49	77.8	75.2	2.6	3
Souttea, Asia.............			81.2	78.5	2.7	1
Puerto d'Espana, S. America .	10 38	61 34	79.5	76.5	3.0	1
Singapore, Asia...........	1 17	—103 50	82.2	78.5	3.7	6
Kingstown, St. Vincent......	13 8	60 37	81.8	78.1	3.7	1
Kandy, Ceylon............	7 17	— 80 49	74.6	70.6	4.0	3
St. Vincent, West Indies.....	13 10	60 31	83.3	79.3	4.0	6
Caraccas, South America....	10 31	67 5	73.5	69.4	4.1	1
Samarang, Java...........	— 6 50	—110 30	84.2	80.1	4.1	1
Bogotá, South America	4 36	74 14	61.9	57.6	4.3	1
Tovar, South America	10 31	67 30	66.0	61.5	4.5	1
Barbadoes, West Indies.....	13 4	59 37	80.6	76.1	4.5	1
St. Bartholomew, West Indies.	17 53	62 54	83.3	78.7	4.6	1
La Guayra, South America...	10 37	67 7	81.1	76.5	4.6	1
Freetown, West Africa	8 30	13 10	82.0	77.0	5.0	1
Batavia, Java.............	— 6 9	—106 53	80.0	75.0	5.0	1
Trevandrum, Hindostan.....	8 31	— 74 50	82.7	77.7	5.0	8
Raiatea, Society Islands.....	—16 40	156 16	80.9	75.7	5.2	1
Antigua, West Indies.......	17 8	61 48	81.9	76.5	5.4	1
Paramaribo, Dutch Guiana...	5 44	55 13	83.4	78.0	5.4	2
Guatemala, Central America..	14 36	90 28	71.2	65.8	5.4	1
St. Thomas, West Indies.....	18 21	64 56	83.7	78.2	5.5	1
Upper Park Camp, Jamaica..	17 58	76 50	83.0	77.5	5.5	1

TABLE XXI.—PLACES HAVING A GREAT MONTHLY RANGE OF TEMPERATURE.

Place.	Latitude.	Longitude.	Hottest Month.	Coldest Month.	Differ- ence.	No. of Years.
	o ′	o ′	o	o	o	
Jakutsk, Siberia...........	62 2	—129 44	62.2	—43.8	106.0	17
Ustjansk, Siberia	70 55	—138 24	52.7	—38.9	91.6	4
Fort Churchill, Hudson Bay..	59 2	93 10	58.0	—28.0	86.0	1
Nertchinsk, Russia.........	51 18	—119 20	64.0	—21.3	85.3	14
Udskoi Ostrog, Siberia......	54 30	—134 58	61.0	—21.6	82.6	1
Utkinsk, Russia	57 45	— 59 22	73.9	— 6.9	80.8	1
Kirgis, Russia.............	50 0	— 60 0	67.5	—13.0	80.5	1
Uralsk, Russia	51 11	— 51 22	78.4	— 0.2	78.6	3
Fort Simpson, British America	62 11	121 32	63.5	—13.5	77.0	2
Cumberland House, Br. Amer.	53 57	102 17	61.8	—13.2	75.0	1
Melville Island, Br. America..	74 47	110 48	42.4	—32.5	74.9	1
Fort Franklin, Great Bear Lake	65 12	123 13	52.1	—22.3	74.4	1
Boothia Felix, Br. America...	69 59	92 1	41.3	—32.0	73.3	3
Barnaul, Russia	53 20	— 83 27	67.5	— 5.6	73.1	6
Nischney Tugilsk, Russia....	57 56	— 60 8	70.3	1.2	69.1	2
Bogoslowsk, Russia	59 45	— 59 59	66.0	— 3.1	69.1	6
Tomsk, Russia	56 30	— 85 10	65.3	— 3.5	68.8	5
Irkutsk, Russia	52 17	—104 17	64.8	— 3.3	68.1	10
Igloolik, British America....	69 21	81 53	39.1	—28.2	67.3	1
Orenburg, Russia	50 46	— 55 6	67.2	0.8	66.4	8

TABLE XXII.—PLACES HAVING SMALL ABSOLUTE RANGE OF TEMPERATURE.

Place.	Latitude.	Longitude.	Highest.	Lowest.	Range.
	° ′	° ′	°	°	°
Barbadoes, West Indies	13 5	59 37	86	72	14
Pulo Penang, Malacca Strait .	5 25	—100 19	90	76	14
Curaçao, South America	12 6	69 20	91	75	16
San Luis de Maranha, Brazil. .	— 2 31	. 44 18	92	76	16
Surinam, Dutch Guiana.	5 38	55 20	90	70	20
La Guayra, Venezuela	10 36	67 7	91	70	21
Cayenne, Guiana	4 56	52 17	87	65	22
Amboyno, E. Archipelago . . .	3 41	—128 17	91	68	23
Tahiti, South Pacific.	—17 29	149 30	90	65	25
Maracaibo, Venezuela.	10 43	71 52	99	70	29
Singapore, Malacca	. 1 17	—103 50	95	66	29
Quito, Equador	— 0 14	78 45	72	43	29
Lima, Peru	—12 3	77 8	86	57	29
St. Helena, South Atlantic . . .	—15 55	5 43	82	52	30
Port Louis, Isle of France . . .	—20 10	— 57 30	91	60	31
Martinique, West Indies.	14 40	61 3	95	63	32
Trinidad, Caribbean Sea	10 39	61 23	93	61	32
St. Bartholomew, West Indies.	17 54	62 54	97	64	33
Paramaribo, Guiana.	5 45	55 13	94	61	33
Funchal, Madeira.	32 38	16 56	85	51	34
Vera Cruz, Mexico.	19 12	96 9	96	61	35
Fort Dundas, Australia	—11 25	—132 25	100·	63	37

TABLE XXIII.—PLACES HAVING LARGE ABSOLUTE RANGE OF TEMPERATURE.

Place.	Latitude.	Longitude.	Highest.	Lowest.	Range.
	° ′	° ′	°	°	°
Barnaul, Asia.	53 20	— 83 27	96	—67	163
Jakutzk, Siberia	62 2	—129 44	86	—76	162
Nijnei Taguilsk, Ural Mts. . . .	57 56	— 60 8	95	—60	155
Bogoslowsk, Ural Mts.	59 45	— 59 59	91	—63	154
Fort Reliance, Brit. America. .	62 46	109 0	81	—70	151
Zlatouste, Ural Mts.	55 11	— 59 45	88	—57	145
Nertchinsk, Siberia	51 18	—119 20	94	—50	144
Catherinenburg, Ural Mts. . . .	56 50	— 60 34	94	—48	142
Moscow, Russia	55 45	— 37 34	94	—47	141
Montreal, Canada.	45 31	. 73 32	102	—38	140
Lowville, New York	43 47	75 33	100	—40	140
Quebec, Canada.	46 49	71 12	99	—40	139
Fort Howard, Wisconsin.	44 30	88 5	100	—38	138
Nijnei Kolymsk, Siberia.	68 32	—160 56	72	—65	137
Enontakis, Lapland	68 30	— 20 47	79	—58	137
Kazan, Russia	55 48	— 49 7	97	—40	137
Fort Snelling, Wisconsin	44 53	93 10	100	—37	137
Montgomery, New York	41 32	74 0	104	—33	137
Tornea, Lapland	66 27	— 23 55	77	—58	135
Lougan, Russia	48 35	— 39 21	101	—33	134
Granville, New York	44 20	73 17	102	—31	133
St. Louis, Missouri	38 37	90 15	108	—25	133
Kinderhook, New York.	42 22	73 43	102	—30	132
Chicago, Illinois	41 53	87 37	102	—30	132
Albany, New York.	42 39	73 44	99	—32	131

TABLE XXIV.

HEIGHT OF THE SNOW LINE ABOVE THE SEA.

Mountains.	Latitude.	Height.	Mountains.	Latitude.	Height.
	° ′	Feet.		° ′	Feet.
Spitzbergen	78 N.	0	Bolor Mountains ...	37 30 N.	17,010
North Cape	71 10	2,400	Hindu Kho	34 30	15,735
Mountains of Norway	70 30	3,178	Himalaya, north side	30	17,392
Suñtelma, Lapland..	67	3,835	Himalaya, south side	28	14,280
Iceland	65	3,080	Cordilleras of ⎫		
Mountains of Norway	62	5,155	Mexico ⎬	19	14,868
Aldan Mount- ⎫ ⎬	60 55	4,470	Mountains of ⎫		
ains, Siberia ⎭			Abyssinia ⎬	13 10	14,065
Kamtschatka	59 30	5,249	Sierra Nevada ⎫		
Mountains of Norway	59 30	5,423	of Merida ⎬	8 5	14,920
Unalaschka, W. ⎫ ⎬ ...	56 30	3,510	Volcano of Tolima..	4 46	15,325
America ⎭			Puraci, S. America..	2 15 N.	15,381
Altai Mountains....	50	7,246			
Alps	45 45	8,890	Nevados of Quito...	0 0	15,960
Caucasus	43 20	10,818	Cotopaxi	0 41 S.	15,924
Rocky Mountains...	43	12,467	Arequipa, Bolivia...	16	17,250
Pyrenees	42 45	8,676	Paachata, Bolivia...	18	18,524
Ararat	39 42	14,170	Portillo, Chili	33	14,708
Mount Argæus	38 33	10,705	Cordilleras, Chili ...	42 30	6,000
Etna	37 30 N.	9,485	Magellan Strait	53 30 S.	3,707

TABLE XXV.

FACTORS FOR MULTIPLYING THE EXCESS OF THE DRY-BULB OVER THE WET-BULB THERMOMETER, TO FIND THE EXCESS OF THE TEMPERATURE OF THE AIR ABOVE THAT OF THE DEW-POINT.

Dry-bulb Therm.	Factor.	Dry-bulb Therm.	Factor.	Dry-bulb Therm.	Factor.	Dry-bulb Therm.	Factor.	Dry-bulb Therm.	Factor.
°		°		°		°		°	
10	8.78	25	6.53	40	2.29	55	1.96	70	1.77	85	1.65
11	8.78	26	6.08	41	2.26	56	1.94	71	1.76	86	1.65
12	8.78	27	5.61	42	2.23	57	1.92	72	1.75	87	1.64
13	8.77	28	5.12	43	2.20	58	1.90	73	1.74	88	1.64
14	8.76	29	4.63	44	2.18	59	1.89	74	1.73	89	1.63
15	8.75	30	4.15	45	2.16	60	1.88	75	1.72	90	1.63
16	8.70	31	3.70	46	2.14	61	1.87	76	1.71	91	1.62
17	8.62	32	3.32	47	2.12	62	1.86	77	1.70	92	1.62
18	8.50	33	3.01	48	2.10	63	1.85	78	1.69	93	1.61
19	8.34	34	2.77	49	2.08	64	1.83	79	1.69	94	1.60
20	8.14	35	2.62	50	2.06	65	1.82	80	1.68	95	1.60
21	7.88	36	2.50	51	2.04	66	1.81	81	1.68	96	1.59
22	7.60	37	2.42	52	2.02	67	1.80	82	1.67	97	1.59
23	7.28	38	2.36	53	2.00	68	1.79	83	1.67	98	1.58
24	6.92	39	2.32	54	1.98	69	1.78	84	1.66	99	1.58

TABLE XXVI.

RELATIVE HUMIDITY OF THE AIR.

Temp. of Air.	Difference of Temperature of the Air and of the Dew Point.																	
	0°	1°	2°	3°	4°	5°	6°	7°	8°	9°	10°	12°	14°	16°	18°	20°	22°	24°
6°	100	96	91	87	83	80	76	72	69	66	63	57	52	47	43	39	35	32
7	100	96	91	87	83	80	76	73	69	66	63	57	52	47	43	39	35	32
8	100	96	91	87	83	80	76	73	69	66	63	57	52	47	43	39	35	·32
9	100	96	91	87	83	80	76	73	69	66	63	58	52	48	43	39	35	32
10	100	96	91	87	83	80	76	73	70	66	63	58	52	48	43	39	36	32
11	100	96	91	87	83	80	76	73	70	66	63	58	53	48	43	39	36	33
12	100	95	91	87	83	80	76	73	70	66	63	58	53	48	44	40	36	33
13	100	95	91	87	83	80	76	73	70	66	63	58	53	48	44	40	36	33
14	100	95	91	87	83	80	76	73	70	66	64	58	53	48	44	40	36	33
15	100	95	91	87	83	80	76	73	70	67	64	58	53	48	44	40	36	33
16	100	95	91	87	83	80	76	73	70	66	64	58	53	48	44	40	36	33
17	100	95	91	87	83	80	76	73	70	66	64	58	53	48	44	40	36	33
18	100	95	91	87	83	80	76	73	69	66	63	58	53	48	44	40	36	33
19	100	95	91	87	83	80	76	73	69	66	63	58	53	48	44	40	37	33
20	100	96	91	87	83	80	76	73	69	66	63	58	53	48	44	40	37	33
21	100	96	91	87	83	80	76	73	69	66	63	58	53	48	44	40	37	33
22	100	96	91	87	83	80	76	73	70	66	63	58	53	48	44	40	37	33
23	100	96	91	87	83	80	76	73	70	66	64	58	53	48	44	40	37	34
24	100	96	91	87	84	80	76	73	70	67	64	58	53	48	44	40	37	34
25	100	96	91	88	84	80	76	73	70	67	64	58	53	48	44	40	37	34
26	100	96	92	88	84	80	77	73	70	67	64	58	53	49	44	40	37	34
27	100	96	92	88	84	80	77	73	70	67	64	58	53	49	44	41	37	34
28	100	96	92	88	84	80	77	74	70	67	64	59	53	49	45	41	37	34
29	100	96	92	88	84	81	77	74	70	67	64	59	54	49	45	41	37	34
30	100	96	92	88	84	81	77	74	71	68	65	59	54	49	45	41	37	34
31	100	96	92	88	84	81	77	74	71	68	65	59	54	49	45	41	37	34
32	100	96	92	88	85	81	78	74	71	68	65	60	54	50	45	41	38	34
33	100	96	92	89	85	81	78	75	71	68	65	60	55	50	46	42	38	35
34	100	96	92	89	85	82	78	75	72	69	66	60	55	50	46	42	38	35
35	100	96	92	89	85	82	78	75	72	69	66	60	55	51	46	42	38	35
36	100	96	92	89	85	82	79	75	72	69	66	61	56	51	46	42	39	35
37	100	96	92	89	85	82	79	76	72	69	67	61	56	51	47	43	39	36
38	100	96	92	89	85	82	79	76	73	70	67	61	56	51	47	43	39	36
39	100	96	92	89	85	82	79	76	73	70	67	62	56	52	47	43	39	36
40	100	96	92	89	86	82	79	76	73	70	67	62	57	52	48	43	40	36
41	100	96	93	89	86	82	79	76	73	70	67	62	57	52	48	44	40	36
42	100	96	93	89	86	82	79	76	73	70	68	62	57	53	48	44	40	37
43	100	96	93	89	86	82	79	76	73	71	68	62	58	53	48	44	41	37
44	100	96	93	89	86	83	79	76	73	71	68	63	58	53	49	45	41	37
45	100	96	93	89	86	83	80	76	74	71	68	63	58	53	49	45	41	38
46	100	96	93	89	86	83	80	77	74	71	68	63	58	54	49	45	41	38
47	100	96	93	89	86	83	80	77	74	71	68	63	58	54	49	45	42	38
48	100	96	93	89	86	83	80	77	74	71	68	63	58	54	50	46	42	38
49	100	96	93	89	86	83	80	77	74	71	68	63	59	54	50	46	42	39
50	100	96	93	89	86	83	80	77	74	71	69	63	59	54	50	46	42	39

TABLE XXVI.—RELATIVE HUMIDITY OF THE AIR.   275

TABLE XXVI.

RELATIVE HUMIDITY OF THE AIR.

Temp. of Air.	Difference of Temperature of the Air and of the Dew Point.																	
	0°	1°	2°	3°	4°	5°	6°	7°	8°	9°	10°	12°	14°	16°	18°	20°	22°	24°
51°	100	96	93	89	86	83	80	77	74	71	69	64	59	54	50	46	43	39
52	100	96	93	89	86	83	80	77	74	71	69	64	59	55	50	47	43	39
53	100	96	93	90	86	83	80	77	74	72	69	64	59	55	51	47	43	40
54	100	96	93	90	86	83	80	77	74	72	69	64	59	55	51	47	43	40
55	100	96	93	90	86	83	80	77	74	72	69	64	59	55	51	47	43	40
56	100	96	93	90	86	83	80	77	75	72	69	64	59	55	51	47	44	40
57	100	96	93	90	86	83	80	77	75	72	69	64	60	55	51	47	44	40
58	100	96	93	90	87	83	80	78	75	72	69	64	60	55	51	47	44	41
59	100	96	93	90	87	84	81	78	75	72	70	65	60	55	51	48	44	41
60	100	96	93	90	87	84	81	78	75	72	70	65	60	56	52	48	44	41
61	100	96	93	90	87	84	81	78	75	72	70	65	60	56	52	48	44	41
62	100	97	93	90	87	84	81	78	75	72	70	65	60	56	52	48	44	41
63	100	97	93	90	87	84	81	78	75	72	70	65	60	56	52	48	45	41
64	100	97	93	90	87	84	81	78	75	73	70	65	60	56	52	48	45	41
65	100	97	93	90	87	84	81	78	75	73	70	65	61	56	52	49	45	42
66	100	97	93	90	87	84	81	78	75	73	70	65	61	56	52	49	45	42
67	100	97	93	90	87	84	81	78	76	73	70	65	61	57	53	49	45	42
68	100	97	93	90	87	84	81	78	76	73	70	66	61	57	53	49	45	42
69	100	97	93	90	87	84	81	78	76	73	71	66	61	57	53	49	46	42
70	100	97	93	90	87	84	81	78	76	73	71	66	61	57	53	49	46	42
71	100	97	93	90	87	84	81	79	76	73	71	66	61	57	53	49	46	43
72	100	97	93	90	87	84	81	79	76	73	71	66	61	57	53	49	46	43
73	100	97	93	90	87	84	81	79	76	73	71	66	62	57	53	50	46	43
74	100	97	93	90	87	84	82	79	76	74	71	66	62	57	53	50	46	43
75	100	97	93	90	87	84	82	79	76	74	71	66	62	58	54	50	46	43
76	100	97	94	90	87	85	82	79	76	74	71	66	62	58	54	50	47	43
77	100	97	94	90	87	85	82	79	76	74	71	67	62	58	54	50	47	43
78	100	97	94	90	88	85	82	79	76	74	71	67	62	58	54	50	47	44
79	100	97	94	91	88	85	82	79	77	74	71	67	62	58	54	50	47	44
80	100	97	94	91	88	85	82	79	77	74	72	67	62	58	54	51	47	44
81	100	97	94	91	88	85	82	79	77	74	72	67	63	58	54	51	47	44
82	100	97	94	91	88	85	82	79	77	74	72	67	63	58	55	51	47	44
83	100	97	94	91	88	85	82	79	77	74	72	67	63	59	55	51	48	44
84	100	97	94	91	88	85	82	79	77	74	72	67	63	59	55	51	48	44
85	100	97	94	91	88	85	82	80	77	75	72	67	63	59	55	51·	48	45
86	100	97	94	91	88	85	82	80	77	75	72	67	63	59	55	51	48	45
87	100	97	94	91	88	85	82	80	77	75	72	68	63	59	55	52	48	45
88	100	97	94	91	88	85	82	80	77	75	72	68	63	59	55	52	48	45
89	100	97	94	91	88	85	83	80	77	75	72	68	63	59	55	52	48	45
90	100	97	94	91	88	85	83	80	77	75	73	68	64	59	56	52	49	45
91	100	97	94	91	88	85	83	80	77	75	73	68	64	60	56	52	49	45
92	100	97	94	91	88	85	83	80	78	75	73	68	64	60	56	52	49	46
93	100	97	94	91	88	85	83	80	78	75	73	68	64	60	56	52	49	46
94	100	97	94	91	88	86	83	80	78	75	73	68	64	60	56	52	49	46
95	100	97	94	91	88	86	83	80	78	75	73	68	64	60	56	53	49	46

TABLE XXVII.

ELASTIC FORCE OF AQUEOUS VAPOR.

Temperature.	Force of Vapor.	Temperature.	Force of Vapor.	Temperature.	Force of Vapor.	Temperature.	Force of Vapor.	Temperature.	Force of Vapor.
°	Inch.	°	Inch.	°	Inch.	°	Inch.	°	Inch.
—30	.009	47	.323	69	.708	81.4	1.071	91.4	1.473
—25	.012	47.5	.329	69.3	.716	81.6	1.078	91.6	1.482
—20	.016	48	.335	69.7	.725	81.8	1.085	91.8	1.491
—15	.021	48.5	.341	70	.733	82.0	1.092	92.0	1.501
—10	.027	49	.348	70.3	.740	82.2	1.099	92.2	1.510
— 5	.034	49.5	.354	70.7	.751	82.4	1.106	92.4	1.520
0	.043	50	.361	71	.758	82.6	1.114	92.6	1.529
+ 2	.048	50.5	.367	71.3	.766	82.8	1.121	92.8	1.539
4	.052	51	.374	71.7	.776	83.0	1.128	93.0	1.548
6	.057	51.5	.381	72	.784	83.2	1.135	93.2	1.558
8	.062	52	.388	72.3	.792	83.4	1.143	93.4	1.568
10	.068	52.5	.395	72.7	.803	83.6	1.150	93.6	1.577
12	.075	53	.403	73	.811	83.8	1.158	93.8	1.587
14	.082	53.5	.410	73.3	.820	84.0	1.165	94.0	1.597
16	.090	54	.418	73.7	.831	84.2	1.173	94.2	1.607
18	.098	54.5	.425	74	.839	84.4	1.180	94.4	1.617
20	.108	55	.433	74.3	.848	84.6	1.188	94.6	1.627
21	.113	55.5	.441	74.7	.859	84.8	1.195	94.8	1.637
22	.118	56	.449	75.0	.868	85.0	1.203	95.0	1.647
23	.123	56.5	.457	75.2	.873	85.2	1.211	95.2	1.657
24	.129	57	.465	75.4	.879	85.4	1.219	95.4	1.667
25	.135	57.5	.474	75.6	.885	85.6	1.226	95.6	1.677
26	.141	58	.482	75.8	.891	85.8	1.234	95.8	1.688
27	.147	58.5	.491	76.0	.897	86.0	1.242	96.0	1.698
28	.153	59	.500	76.2	.903	86.2	1.250	96.2	1.708
29	.160	59.5	.509	76.4	.909	86.4	1.258	96.4	1.719
30	.167	60	.518	76.6	.915	86.6	1.266	96.6	1.729
31	.174	60.5	.527	76.8	.921	86.8	1.274	96.8	1.740
32	.181	61	.536	77.0	.927	87.0	1.282	97.0	1.751
33	.188	61.5	.546	77.2	.933	87.2	1.290	97.2	1.761
34	.196	62	.556	77.4	.939	87.4	1.298	97.4	1.772
35	.204	62.5	.566	77.6	.946	87.6	1.307	97.6	1.783
36	.212	63	.576	77.8	.952	87.8	1.315	97.8	1.794
37	.220	63.3	.582	78.0	.958	88.0	1.323	98.0	1.805
38	.229	63.7	.590	78.2	.964	88.2	1.332	98.2	1.816
39	.238	64	.596	78.4	.971	88.4	1.340	98.4	1.827
40	.248	64.3	.602	78.6	.977	88.6	1.349	98.6	1.838
40.5	.252	64.7	.611	78.8	.984	88.8	1.357	98.8	1.849
41	.257	65	.617	79.0	.990	89.0	1.366	99.0	1.861
41.5	.262	65.3	.624	79.2	.997	89.2	1.375	99.2	1.872
42	.267	65.7	.632	79.4	1.003	89.4	1.383	99.4	1.883
42.5	.272	66	.639	79.6	1.010	89.6	1.392	99.6	1.895
43	.277	66.3	.646	79.8	1.016	89.8	1.401	99.8	1.906
43.5	.283	66.7	.655	80.0	1.023	90.0	1.410	100.0	1.918
44	.288	67	.662	80.2	1.030	90.2	1.419	100.2	1.929
44.5	.294	67.3	.668	80.4	1.037	90.4	1.427	100.4	1.941
45	.299	67.7	.678	80.6	1.043	90.6	1.436	100.6	1.953
45.5	.305	68	.685	80.8	1.050	90.8	1.446	100.8	1.965
46	.311	68.3	.692	81.0	1.057	91.0	1.455	101.0	1.977
46.5	.317	68.7	.701	81.2	1.064	91.2	1.464	101.2	1.988

TABLE XXVIII.—PRESSURE AND VELOCITY OF THE WIND. 277

TABLE XXVIII.

FOR COMPARING THE PRESSURE AND VELOCITY OF THE WIND.

Pressure, oz. per sq. foot.	Velocity, miles per hour.	Pressure, lbs. per sq. foot.	Velocity, miles per hour.	Pressure, lbs. per sq. foot.	Velocity, miles per hour.	Pressure, lbs. per sq. foot.	Velocity, miles per hour.	Pressure, lbs. per sq. foot.	Velocity, miles per hour.
0.08	1.000	6.75	36.742	17.50	59.160	28.25	75.166	39.25	88.600
0.25	1.767	7.00	37.416	17.75	59.581	28.50	75.498	39.50	88.881
0.50	2.500	7.25	38.078	18.00	60.000	28.75	75.828	39.75	89.162
0.75	3.061	7.50	38.729	18.25	60.415	29.00	76.157	40.00	89.442
1.00	3.535	7.75	39.370	18.50	60.827	29.25	76.485	40.25	89.721
2	5.000	8.00	40.000	18.75	61.237	29.50	76.811	40.50	90.000
3	6.123	8.25	40.620	19.00	61.644	29.75	77.136	40.75	90.277
4	7.071	8.50	41.231	19.25	62.048	30.00	77.459	41.00	90.553
5	7.905	8.75	41.833	19.50	62.449	30.25	77.781	41.25	90.829
6	8.660	9.00	42.426	19.75	62.849	30.50	78.102	41.50	91.104
7	9.354	9.25	43.011	20.00	63.245	30.75	78.421	41.75	91.378
8	10.000	9.50	43.588	20.25	63.639	31.00	78.740	42.00	91.651
9	10.606	9.75	44.158	20.50	64.031	31.25	79.056	42.25	91.923
10	11.180	10.00	44.721	20.75	64.420	31.50	79.372	42.50	92.195
11	11.726	10.25	45.276	21.00	64.807	31.75	79.686	42.75	92.466
12	12.247	10.50	45.825	21.25	65.192	32.00	80.000	43.00	92.736
13	12.747	10.75	46.368	21.50	65.574	32.25	80.311	43.25	93.005
14	13.228	11.00	46.904	21.75	65.954	32.50	80.622	43.50	93.273
15	13.693	11.25	47.434	22.00	66.332	32.75	80.932	43.75	93.541
Pounds.		11.50	47.958	22.25	66.708	33.00	81.240	44.00	93.808
1	14.142	11.75	48.476	22.50	67.082	33.25	81.547	44.25	94.074
1.25	15.811	12.00	48.989	22.75	67.453	33.50	81.853	44.50	94.339
1.50	17.320	12.25	49.497	23.00	67.823	33.75	82.158	44.75	94.604
1.75	18.708	12.50	50.000	23.25	68.190	34.00	82.462	45.00	94.868
2.00	20.000	12.75	50.497	23.50	68.556	34.25	82.764	45.25	95.131
2.25	21.213	13.00	50.990	23.75	68.920	34.50	83.066	45.50	95.393
2.50	22.360	13.25	51.478	24.00	69.282	34.75	83.366	45.75	95.655
2.75	23.452	13.50	51.961	24.25	69.641	35.00	83.666	46.00	95.916
3.00	24.494	13.75	52.440	24.50	70.000	35.25	83.964	46.25	96.176
3.25	25.495	14.00	52.915	24.75	70.356	35.50	84.261	46.50	96.436
3.50	26.457	14.25	53.385	25.00	70.710	35.75	84.557	46.75	96.695
3.75	27.386	14.50	53.851	25.25	71.063	36.00	84.852	47.00	96.953
4.00	28.284	14.75	54.313	25.50	71.414	36.25	85.146	47.25	97.211
4.25	29.154	15.00	54.772	25.75	71.763	36.50	85.440	47.50	97.467
4.50	30.000	15.25	55.226	26.00	72.111	36.75	85.732	47.75	97.724
4.75	30.822	15.50	55.677	26.25	72.456	37.00	86.023	48.00	97.979
5.00	31.622	15.75	56.124	26.50	72.801	37.25	86.313	48.25	98.234
5.25	32.403	16.00	56.568	26.75	73.143	37.50	86.602	48.50	98.488
5.50	33.166	16.25	57.008	27.00	73.484	37.75	86.890	48.75	98.742
5.75	33.911	16.50	57.445	27.25	73.824	38.00	87.177	49.00	98.994
6.00	34.641	16.75	57.879	27.50	74.161	38.25	87.464	49.25	99.247
6.25	35.355	17.00	58.309	27.75	74.498	38.50	87.749	49.50	99.498
6.50	36.055	17.25	58.736	28.00	74.833	38.75	88.034	49.75	99.749
6.75	36.742	17.50	59.160	28.25	75.166	39.00	88.317	50.00	100.000

TABLE XXIX.

AVERAGE AMOUNT OF RAIN FOR EACH MONTH, SEASON, AND THE YEAR.

Station.	Lat.		Long.		Alt.	Jan.	Feb.	March.	April.	May.
	°	′	°	′	Feet.	Inches.	Inches.	Inches.	Inches.	Inches.
Paramaribo, Dutch Guiana ..	5	44	55	13		18.74	16.54	20.75	21.10	23.23
Caraccas, Venezuela........	10	22	67	12		1.00	0.25	1.10	1.20	17.00
Matouba, Guadeloupe......	16	10	61	50		21.30	17.76	22.64	21.38	18.11
Vera Cruz, Mexico.........	19	12	96	9	50	5.10	0.00	0.00	0.50	31.40
Havana, Cuba.............	23	9	82	23	50	4.97	3.08	4.08	2.28	10.11
Key West, Florida.........	24	32	81	48	10	2.20	1.22	.2.83	1.34	3.92
Corpus Christi, Texas	27	47	97	27	20	3.96	2.37	1.25	4.01	4.68
Fort Brooke, Texas........	28	0	82	28	20	2.20	3.01	3.37	1.95	3.24
St. Augustine, Florida......	29	48	81	35	25	2.09	1.63	2.34	1.56	2.00
New Orleans, Louisiana.....	29	57	90	0	10	5.61	2.90	3.90	3.29	4.10
Mobile, Alabama..........	30	42	88	1	30	8.89	5.07	5.86	4.95	3.43
Savannah, Georgia........	32	6	81	5	30	2.76	2.53	3.69	2.11	5.20
San Diego, California.......	32	42	117	13	150	0.83	2.01	1.40	0.77	0.57
Charleston, South Carolina...	32	46	79	56	30	2.33	3.39	3.02	1.72	3.66
Santa Fé, New Mexico......	35	41	106	1	6846	0.31	0.57	1.29	0.80	0.74
Nashville, Tennessee	36	9	86	49	533	5.01	3.98	4.91	5.20	4.94
Norfolk, Virginia..........	36	50	76	19	8	3.26	2.74	3.33	2.80	3.64
Fort Massachusétts, New Mex.	37	·32	105	23	8365	0.23	·0.72	0.94	0.42	2.14
San Francisco, California....	37	48	122	27	150	3.23	3.31	4.61	3.72	0.48
Sacramento, California......	38	35	121	28	50	2.98	2.36	3.97	1.44	0.87
St. Louis, Missouri........	38	37	90	15	481	2.03	2.23	3.40	3.93	4.97
Washington, D. C..........	38	53	77	0	78	4.45	2.75	2.57	4.03	3.85
Cincinnati, Ohio..........	39	6	84	30	550	3.35	3.51	3.93	3.66	4.55
Philadelphia, Pennsylvania ..	39	57	75	10	30	3.09	2.94	3.43	3.64	3.90
Pittsburg, Pennsylvania.....	40	32	80	2	704	2.18	2.17	2.70	3.10	3.58
New York City, New York ..	40	43	74	0	23	2.78	2.92	3.44	3.33	4.78
Salt Lake City, Utah	40	46	112	6	4351	1.23	1.99	2.34	1.66	1.34
New Haven, Connecticut....	41	18	72	55	60	5.53	3.97	3.49	3.31	4.23
Fort Laramie, Dacotah......	42	12	104	48	4519	0.27	0.71	1.37	1.93	5.39
Detroit, Michigan.........	42	20	83	2	580	2.18	1.38	2.86	2.92	2.73
Boston, Massachusetts	42	21	71	3	71	2.39	3.19	3.47	3.64	3.74
Albany, New York	42	40	73	45	130	2.77	2.62	2.82	3.12	3.85
Fort Orford, Oregon	42	44	124	29	50	8.81	6.35	8.24	5.64	5.24
Milwaukee, Wisconsin	43	4	87	54	593	1.30	0.80	1.60	2.40	2.50
Rochester, New York.......	43	8	77	51	506	1.88	1.40	1.81	1.97	3.04
Toronto, Canada..........	43	39	79	23	341	1.70	1.09	1.61	2.57	2.98
Fort Snelling, Wisconsin....	44	53	93	10	820	0.73	0.52	1.30	2.14	3.17
Wolfville, Nova Scotia......	45	6	64	25		3.27	2.54	5.94	4.20	3.30
Montreal, Canada	45	31	73	33		3.21	4.19	3.32	2.48	2.82
Astoria, Oregon...........	46	11	123	18	50	27.00	10.95	6.10	4.38	5.95
Fort Brady, Michigan......	46	39	84	43	600	1.13	1.37		1.83	2.24
Steilacoom, Washington Ter..	47	10	122	25	300	9.54	5.16	4.56	4.77	1.86
Fort Kent, Maine..........	47	15	68	35	575	3.73	2.60	1.77	1.06	2.63
St. Johns, Newfoundland....	47	33	52	28	140	4.74	2.75	4.80	3.76	4.13
Sitka, Aliashka	57	3	135	18	20	7.80	7.32	6.20	6.83	5.29

TABLE XXIX.

AVERAGE AMOUNT OF RAIN FOR EACH MONTH, SEASON, AND THE YEAR.

June.	July.	Aug.	Sept.	Oct.	Nov.	Dec.	Spring.	Summer.	Autumn.	Winter.	Year.
Inches.	Inches.	Inches.	Inches.	Inches.	Inches.	Inches.	Inches.	Inches.	Inches.	Inches.	Inches.
16.34	5.89	1.77	0.63	1.46	2.99	13.03	65.08	24.00	5.08	48.31	142.47
16.00	14.04	21.14	39.37	13.40	26.80	4.07	19.30	51.18	79.57	5.32	155.37
39.53	27.95	10.20	13.15	33.11	24.13	43.07	62.13	77.68	70.39	82.13	292.33
21.20	59.70	35.90	38.90	8.00	4.50	0.40	31.90	116.80	51.40	5.50	183.20
25.28	5.93	6.90	11.14	11.01	4.74	1.83	16.47	38.11	26.89	9.88	91.35
5.48	2.97	4.33	6.12	4.94	1.77	2.09	8.09	12.78	12.83	5.51	39.21
5.63	4.89	2.91	6.73	2.37	1.05	1.26	9.94	13.43	10.15	7.59	41.11
7.04	11.10	10.10	6.23	2.40	2.00	2.83	8.56	28.24	10.63	8.04	55.47
4.27	3.24	3.03	5.85	2.42	1.29	2.08	5.90	10.54	9.56	5.80	31.80
4.97	6.66	5.65	2.20	2.74	4.68	4.20	11.29	17.28	9.62	12.71	50.90
5.05	4.36	8.59	4.68	2.65	6.58	4.31	14.24	18.00	13.91	18.27	64.42
4.84	7.57	8.32	4.26	2.55	1.65	3.20	11.00	20.72	8.46	8.48	48.66
0.15	0.01	0.39	0.03	0.05	1.16	3.06	2.74	0.55	1.24	5.90	10.43
5.00	6.15	7.53	6.34	3.04	2.23	3.68	8.60	18.68	11.61	9.40	48.29
1.32	4.18	3.40	2.55	1.60	1.87	1.20	2.83	8.90	6.02	2.08	19.83
4.41	3.84	4.40	4.94	3.68	3.92	2.96	14.10	14.00	12.30	12.40	52.80
3.78	5.56	5.70	3.93	2.82	3.41	4.17	9.77	15.08	10.16	10.17	45.18
0.74	2.59	2.05	1.39	1.10	6.34	1.88	3.50	5.38	8.83	2.83	20.54
0.02	0.00	0.01	0.07	0.63	2.05	4.71	8.81	0.03	2.75	11.25	22.84
0.09	0.11	0.00	0.01	0.42	3.18	4.42	6.28	0.20	3.61	9.76	19.85
6.06	3.86	4.22	2.57	3.29	3.08	2.68	12.30	14.14	8.94	6.94	42.32
2.93	3.92	3.67	3.52	3.55	3.09	2.87	10.45	10.52	10.16	10.07	41.20
5.01	4.37	4.32	3.10	3.32	3.48	4.29	12.14	13.70	9.90	11.15	46.89
3.57	4.22	4.67	3.53	3.18	3.36	4.03	10.97	12.46	10.07	10.06	43.56
3.56	2.97	3.34	2.68	2.87	2.68	3.13	9.38	9.87	8.23	7.48	34.96
3.46	3.17	4.70	3.31	3.40	3.59	3.93	11.55	11.33	10.30	9.63	42.81
2.15	3.99	0.64	0.85	1.57	3.85	3.68	5.34	6.78	6.27	6.90	25.29
3.30	4.19	4.14	3.88	3.60	3.72	3.43	11.03	11.63	11.20	10.93	44.79
2.95	1.83	0.92	1.33	1.26	1.37	0.65	8.69	5.70	3.96	1.63	19.98
3.91	3.20	2.18	3.31	2.04	2.06	1.30	8.51	9.29	7.41	4.86	30.07
3.13	2.57	5.47	4.27	3.73	4.57	4.31	10.85	11.17	12.57	9.89	44.48
4.48	4.39	3.44	3.34	3.69	3.24	2.91	9.79	12.31	10.27	8.30	40.67
1.06	0.16	1.78	2.34	7.31	10.27	14.43	19.12	3.00	19.92	29.59	71.63
4.00	3.00	2.80	3.20	1.40	2.10	2.00	6.50	9.70	6.80	4.20	27.20
3.25	3.01	2.60	3.05	3.39	2.94	2.10	6.82	8.86	9.38	5.38	30.44
3.04	3.72	2.81	4.46	2.96	2.91	1.50	7.16	9.57	10.33	4.29	31.35
3.63	4.11	3.18	3.32	1.35	1.31	0.67	6.61	10.92	5.98	1.92	25.43
4.82	3.27	5.04	3.70	3.66	3.04	3.67	13.44	12.13	10.40	9.48	45.45
2.65	3.27	3.52	3.53	3.87	3.95	4.41	8.62	9.44	11.35	11.81	41.22
2.85	0.00	1.15	1.87	6.70	13.20	6.20	16.43	4.00	21.77	44.15	86.35
2.83	3.75	3.39	4.33	3.35	3.08	2.21	5.44	9.97	10.76	5.18	31.35
1.97	0.34	1.54	2.67	4.43	8.73	7.92	11.19	3.85	15.83	22.62	53.49
1.36	7.72	2.57	1.36	4.41	3.86	3.36	5.46	11.65	9.64	9.71	36.46
5.67	3.82	5.09	5.79	7.88	3.25	5.25	12.69	14.58	16.92	12.74	56.93
3.79	4.15	7.81	11.27	12.32	8.51	8.65	18.32	15.75	32.10	23.77	89.94

TABLE XXX.—PLACES HAVING A SMALL ANNUAL FALL OF RAIN.

Places.	Latitude.		Longitude.		Height.	Amount.	Authority.
	o	′	o	′	Feet.	Inches.	
Lima, Peru	−12	0	77	2	530	0	Arago's Met. Ess., p. 109.
Thebes, Egypt	25	43	32	35		0	Wilk'n's Egypt., v. 4, p. 10.
Near Mourzouk, Fezzan	25	54	— 14	12		0	Gehler, v. 7, p. 1251.
Tatta, North Africa	28	39	6	46		0	Gehler, v. 7, p. 1251.
Cairo, Egypt	30	2	— 31	15		1.31	Arago Melanges, p. 463.
Kurrachee, Hindostan	24	50	— 67	0		1.50	Ph. Trans. 1850, p. 360.
Kotree, Hindostan	25	20	— 68	14		1.74	Ph. Trans. 1850, p. 361.
Biscara, Algeria	34	51	— 5	40	350	2.50	An. Met. 1854, p. 297.
Fort Yuma, California	32	43	114	36	120	3.24	Army Reg., p. 675.
Astrachan, Russia	46	21	— 48	5	70	4.08	Dove Beiträge, p. 183.
Hyderabad, Hindostan	25	20	— 68	20		4.50	Johnston's Ph. Atlas.
Raimsk, Russia	46	4	— 61	47		5.99	Dove Beiträge, p. 183.
Aralich, Russia	39	53	— 44	33	2600	6.15	Dove Beiträge, p. 138.
Mendosa, La Plata	−32	52	69	6	2600	6.50	Zeitsch. für Erd'e 1858, p. 9.
Novo Petrowsk, Russia	44	27	— 50	8	115	6.72	Dove Beiträge, p. 183.
Fort Conrad, New Mex.	33	34	107	9	4576	6.76	Army Reg., p. 673.
San Louis Rey, Cal.	33	13	117	30	20	6.95	Army Reg.
Barnaoul, Siberia	53	20	— 83	27	400	7.47	Kupffer's Annales.
Taos, New Mexico	36	21	105	42	8000	7.48	Army Reg.
Cumana, Venezuela	10	27	64	15		7.52	Gehler, v. 7, p. 1311.
Sevastopol, Russia	44	36	— 33	32		7.67	Heis Wochen't 1866, p. 325.
Socorro, New Mexico	34	10	·106	54	4560	7.86	Army Reg., p. 673.
Sympheropol, Russia	44	57	— 34	6	780	8.75	Heis Wochen't 1858, p. 176.
Bacou, Russia	40	22	— 49	47	−53	9.05	Kupffer's Annales.
Fort Fillmore, New Mex.	32	13	106	42	3937	9.23	Army Reg., p. 673.
Albuquerque, New Mex.	35	6	106	38	5032	9.42	Army Reg., p. 673.

TABLE XXXI.—PLACES HAVING A GREAT ANNUAL FALL OF RAIN.

Places.	Latitude.		Longitude.		Height.	Amount.	Authority.
	o	′	o	′	Feet.	Inches.	
Cherapoonjee, Hindost'n	25	14	— 91	40	4125	592	Herschel's Met., p. 110.
Matouba, Guadeloupe	16	10	61	50	4000 ?	292	Com. Rend., v. 7, p. 743.
Maranhao, Brazil	— 2	31	44	18		280	Gehler, v. 7, p. 1314.
Uttray Mullay, Hindost'n	8	39	— 77	0	4500	267	Ph. Trans. 1850, p. 358.
Mahabalishwar, Hindo'n	17	54	— 73	38	4300	254	Ph. Trans. 1850, p. 354.
Sylket, Hindostan	24	53	— 91	47		209	Br. As. 1852, p. 257.
Stye, England	54		3		1600	206	Buchan's Met., p. 118.
Aracan, Hindostan	20	47	— 93	25		200	Buchan's Met., p. 117.
Augusta Peak, Hindost'n	8		— 77		6200	194	Ph. Trans. 1850, p. 362.
Sierra Leone, W. Africa	8	20	13	8		189	Gehler, v. 7, p. 1314.
Sindola, Hindostan	17		— 73		4600	185	Ph. Trans. 1850, p. 354.
Vera Cruz, Mexico	19	12	96	9		183	Mayer's Mexico.
Sandoway, Hindostan	18	25	— 94	30		178	Br. As. 1852, p. 257.
Maulmein, Birmah	16	3	— 97	38		175	Johnston's Ph. Atlas.
Attaghery, Hindostan	8		— 77		2200	170	Ph. Trans. 1850, p. 362.
St. Benoit, Isl. of Bourbon	−20	51	— 55	30		162	Dove Beiträge, p. 102.
Marmato, New Granada	4	40	74	42	4678	162	Br. As. 1840, p. 116.
Demerara, Guiana	6	45	58	2		156	Berghaus's Atlas.
Caraccas, Colombia	10	22	67	5	2730	155	Dove Beiträge, p. 90.
Akyab, Hindostan	20	8	— 92	52		155	Br. As. 1852, p. 257. [368.
Leogane, St. Domingo	18	30	72	30		150	Malte Brun's Geog., v. 1, p.
Buitenzorg, Java	— 6	37	−106	49		147	Dove Beiträge, p. 102.

TABLE XXXII.

COMPARATIVE RADIATING POWER OF DIFFERENT SUBSTANCES AT NIGHT.

Hare-skin	1316	Copper	839
Rabbit-skin	1240	Charcoal in Powder	776
White raw Wool on Grass	1222	Wood	773
Flax on Grass	1186	Blackened Tin	770
Raw Silk	1107	Lead	757
Unwrought white Cotton Wool	1085	Black-lead in Powder	697
		Zinc	681
Yellow Cotton	1005	Iron	642
Long Grass	1000	Paper	614
Black Wadding on Grass	993	Sawdust	610
Lampblack in Powder	961	Slate	573
Flannel	886	Garden-mould	472
Light blue Lamb's Wool	876	Tin-foil	470
Grass less than an Inch in Height	870	River Sand	454
		Stone	390
Glass	864	Brick	372
Chalk in Powder	840	Gravel	288

TABLE XXXIII.

FALL OF THE BAROMETER IN HURRICANES.

Locality.	Date.	Fall in Inches.	Hours.	Authority.
Near Calcutta	1833, May 21.	2.59	3	Reid's Law of Storms, p. 271.
Bay of Bengal	1840, Apr. 28.	2.05	14	Journal Bengal Soc., v. 9, p. 1014.
South Indian Ocean	1840, May 4.	2.00		Piddington's Horn-Book, p. 215.
St. Thomas, W. I.	1837, Aug. 2.	1.69	6	Poggendorff's Annal., v. 52, p. 25.
Near Calcutta	1832, Oct. 7.	1.60	12	Reid's Law of Storms, p. 269.
Near Hong Kong	1867, Sept. 8.	1.57	13	U. S. Steamer Monocacy.
Bay of Bengal	1852, May 14.	1.55	8	Jour. Bengal Soc., 1855, p. 429.
Bay of Bengal	1854, Apr. 22.	1.50	12	Jour. Bengal Soc., 1858, p. 179.
China Sea	1845, Oct. 9.	1.50	13	Jour. Bengal Soc., v. 18, p. 16.
Mauritius	1818, Feb. 28.	1.50	17	Reid's Law of Storms, p. 141.
Havana, Cuba	1846, Oct. 11.	1.47	6	Piddington's Horn-Book, p. 193.
Macao, China	1832, Aug. 3.	1.46	9	American Journal, v. 35, p. 217.
Calcutta	1842, June 3.	1.42	18	Jour. Bengal Soc., 1842, p. 1004.
Bay of Bengal	1851, Oct. 22.	1.40	7	Jour. Bengal Soc., 1854, p. 513.
Aberdeen, Scotland	1839, Jan. 7.	1.40	12	Espy's Phil. of Storms, p. 521.
Cape Hatteras	1853, Sept. 7.	1.35	7	Amer. Journal, v. 18, N. S., p. 9.
Boston, Mass.	1866, Dec. 27.	1.26	17	R. T. Paine's Journal.
China Sea	1809, Sept. 28.	1.20	12	Jour. Bengal Soc., v. 11, p. 627.
Macao, China	1835, Aug. 5.	1.15	8½	American Journal, v. 35, p. 211.
Chittagong, India.	1849, May 13.	1.06	2¼	Jour. Bengal Soc., 1854, p. 22.
Mauritius	1824, Feb. 23.	1.05	5	Reid's Law of Storms, p. 147.

TABLE XXXIV.

AURORAS, SOLAR SPOTS, AND VARIATION OF THE MAGNETIC NEEDLE.

Year.	Auroras	Year.	Auroras Europe	Auroras America	Solar Spots	Year.	Auroras Europe	Auroras America	Solar Spots	Mag. Var.	Year.	Auroras Europe	Auroras America	Solar Spots	Mag. Var.
1685	1	1740	2			1782	29	24	33	8	1824	0	0	7	8
1686	4	1741	21			1783	17	22	22	9	1825	1	2	17	10
1692	2	1742	14	2		1784	7	4	4		1826	2	0	29	10
1693	2	1743	9	2		1785	14	9	18	8	1827	10	7	40	11
1694	2	1744	8	0		1786	40	55	61	14	1828	11	6	52	12
1695	4	1745	3	0		1787	10	47	93	15	1829	18	2	53	14
1696	4	1746	1	7		1788	10	38	91	13	1830	32	6	59	12
1697	1	1747	7	10		1789	15	51	85	13	1831	23	2	39	12
1698	9	1748	3	6		1790	4	13	75	15	1832	5	2	22	
1699	40	1749	3	10	64	1791	4	12	46	12	1833	12	3	7	
1702	1	1750	12	17	68	1792	1	6	53	9	1834	2	9	11	8
1704	1	1751	2	5	41	1793	2	8	21	8	1835	6	6	45	10
1707	12	1752		2	33	1794	2	2	24	8	1836	8	5	97	12
1708	1	1753		1	23	1795	2	2	16	7	1837	25	41	111	12
1709	3	1754		0	74	1796	1	0	9	8	1838	28	39	83	13
1710	1	1755	1	0	6	1797	1	0	6	8	1839	30	47	68	11
1711	1	1756	2	0	9	1798	0	0	3	7	1840	40	44	52	9
1714	1	1757	0	6	30	1799	2	0	6	7	1841	35	42	30	7
1716	11	1758	2	4	38	1800	3	0	10	7	1842	49	11	19	6
1717	12	1759	8	5	48	1801	4	0	31	8	1843	38	10	9	7
1718	27	1760	7	6	49	1802	4	2	38	8	1844	22	10	13	6
1719	32	1761	12	5	75	1803	6	5	50	9	1845	18	22	33	7
1720	28	1762	18	7	51	1804	6	4	70	8	1846	39	30	47	8
1721	19	1763	4	6	37	1805	4	4	50	9	1847	38	22	79	9
1722	46	1764	9	12	34	1806	3	4	30		1848	38	53	100	11
1723	30	1765	8	7	23	1807	0	2	10		1849	42	20	96	10
1724	26	1766	0	0	17	1808	1	0	2		1850	25	30	64	10
1725	30	1767	5	4	34	1809	0	2	1		1851	17	21	62	8
1726	46	1768	2	7	52	1810	0	0	0		1852	45	42	52	8
1727	67	1769	10	18	86	1811	0	0	1		1853	26	22	38	7
1728	86	1770	13	14	79	1812	0	0	5		1854	36	15	19	·7
1729	65	1771	29	15	73	1813	0	0		7	1855	20		7	6
1730	116	1772	21	7	49	1814	4	3	20	8	1856	20		4	6
1731	57	1773	31	17	40	1815	0	1	35	8	1857	15		22	7
1732	100	1774	48	20	48	1816	1	0	45		1858	34		51	7
1733	27	1775	21	5	27	1817	1	0	44	9	1859	46		96	10
1734	38	1776	12	4	35	1818	2	4	34	9	1860	33		99	10
1735	51	1777	26	15	63	1819	3	6	22	8	1861	35		77	9
1736	43	1778	30	18	95	1820	2	2	9	8	1862	33		59	8
1737	40	1779	37	4	99	1821	2	0	4	9	1863	36		44	8
1738	9	1780	20	25	73	1822	0	1	3	9	1864	47		46	8
1739	27	1781	29	25	68	1823	0	0	1	8	1865	98		30	7

TABLE XXXV.—CATALOGUE OF LARGEST IRON METEORS. 283

TABLE XXXV.

CATALOGUE OF THE LARGEST IRON METEORS.

Locality.	Year found.	Pounds' Weight.	Spec. Grav.	Remarks.
Durango, Mexico......	1811	35,000	7.88	Specimens at Berlin, Vienna, etc.
Otumpa, Buenos Ayres..	1784	33,000	7.60	{ Specimen of 1400 lbs. belongs to British Museum.
Rogue River, Oregon....	1859	22,000		Specimens at Vienna, Boston, etc.
Bemdego River, Brazil..	1784	17,300	7.73	Specimens at Munich, London, etc.
Bonanza, Mexico......	1865	sev'l tons	7.82	Spec. belongs to Prof. C. U. Shepard.
Near Melbourne, Aus. ...	1861	8,287	7.51	Belongs to British Museum.
Sierra Blanca, Mexico ..	1784	4,000	6.50	Specimen at Berlin.
Bitberg, Prussia.......	1802	3,400	6.33	Specimens at Vienna, Berlin, etc.
Near Melbourne, Aus. ...	1861	2,800	7.51	Belongs to Colonial Government.
Zacatecas, Mexico	1792	2,000	7.50	Specimens in Br. Museum, Berlin, etc.
Cocke Co., Tennessee....	1840	2,000	7.26	Belongs to British Museum.
Santa Rosas, New Gran.	1810	1,700	7.30	Specimens at Vienna, Paris, etc.
Jenisey River, Siberia ∴.	1772	1,680	6.48	Belongs to Imp. Mus., St. Petersburg.
Red River, Texas......	1808	1,635	7.70	Belongs to Yale College.
Tucson, Arizona, U. S....	1735	1,400		Belongs to Smithsonian Institute.
La Caille, France......	1828	1,100	7.64	Belongs to Jardin des Plantes, Paris.
Tucson, Arizona, U. S. ...	1863	632	7.29	Belongs to the City of San Francisco.
Tula, Russia..........	1846	542	5.97	Specimens at Vienna, London, etc.
Bear Creek, Colorado ..	1866	436	7.69	Belongs to Prof. C. U. Shepard.
Madoc, Upper Canada..	1854	368		Geological Cabinet at Montreal.
Orange River, S. Africa .	1856	326		Belongs to Prof. C. U. Shepard.
Cape of Good Hope....	1793	300	7.40	Belongs to Haarlem Cabinet, Holland.
Atacama, Bolivia......	1827	300	7.44	Belongs mostly to British Museum.
Pittsburg, Pennsylvania.	1850	292	7.38	Specs. belong to Prof. Silliman, et al.
Carthage, Tennessee....	1846	280		A large spec. belongs to Brit. Museum.
Coahuila, Mexico......	1855	252	7.81	Belongs to Smithsonian Institute.
Seeläsgen, Silesia......	1847	218	7.70	Belongs partly to British Museum.
Toluca, Mexico	1784	218	7.38	150 lbs. belong to Prof. C. U. Shepard.
Brahin, Russia........	1810	200	6.20	Belongs to University at Kiew.
Lenarto, Hungary	1814	194	7.75	Belongs to Museum of Pesth.
Elbogen, Bohemia.....	1811	191	7.74	Chiefly in the Cabinet at Vienna.
Lion River, South Africa	1853	172	7.60	Belongs to Prof. C. U. Shepard.
Walker Co., Alabama ..	1832	165	7.26	Half belongs to British Museum.
Nelson Co., Kentucky ..	1856	161		Specimens in Berlin, London, etc.
Burlington, New York..	1819	150	7.50	Spec. belongs to Prof. C. U. Shepard.
Ruff's Mountain, South Carolina.........	1850	116	7.10	Mostly belongs to Prof. C. U. Shepard.
Lagrange, Oldham Co., Kentucky........	1860	112	7.89	Mostly belongs to Prof. C. U. Shepard.
Bohumilitz, Bohemia...	1829	103	7.60	Belongs to Museum of Prague.
Agram, Croatia	1751	87	7.82	Belongs chiefly to Vienna Cabinet.
Braunau, Silesia.......	1847	72	7.71	Specimens in Vienna, Berlin, etc.
Putnam Co., Georgia...	1839	70	7.69	Belongs partly to Prof. C. U. Shepard.
Tazewell, Claiborne Co., Tennessee	1853	55	7.88	Specimens in London, Berlin, etc.
Schwetz, Prussia......	1850	43	7.77	Chiefly in Berlin.
Denton Co., Texas.....	1856	40	7.67	In Geological Cabinet at Austin.
Claiborne, Clarke Co., Alabama	1834	40	6.50	Spec. belongs to C. T. Jackson, Boston.

TABLE XXXVI.

AEROLITES FALLEN IN THE UNITED STATES.

Locality.	Date of Fall.	Weight in Pounds.	Specific Gravity.	Present Owners.
Weston, Conn. . . .	1807, Dec. 14.	300	3.58	Yale College, et al.
Caswell Co., N. C. .	1810, Jan. 30.	3	3.5 ?	Unknown.
Nobleborough, Me.	1823, Aug. 7.	5	3.09	C. U. Shepard, et al. [al.
Nanjemoy, Md. . . .	1825, Feb. 10.	16	3.66	Yale Coll.; C. U. Shepard, et
Sumner Co., Tenn.	1827, May 9.	11	3.55	C. U. Shepard; Leyden Cab.,
Richmond, Va. . . .	1828, June 4.	4	3.34	C. U. Shepard, et al. [et al.
Forsyth, Monroe Co., Ga.	1829, May 8.	36	3.46	Yale Coll.; C. U. Shepard, et [al.
Deal, N. J.	1829, Aug. 15.	$\frac{1}{16}$	3.25	C. U. Shepard, et al.
Dickson Co., Tenn.	1835, July 31.	9	7.76	Mobile Cabinet, et al.
Little Piney, Mo. .	1839, Feb. 13.	50	3.5	C. U. Shepard, et al.
Bishopville, S. C. .	1843, Mar. 25.	13	3.04	C. U. Shepard, et al. [al.
Linn Co., Iowa . . .	1847, Feb. 25.	75	3.58	Yale Coll.; C. U. Shepard, et
Castine, Me.	1848, May 20.	$\frac{1}{2}$	3.45	Bowdoin College, et al.
Cabarrus Co., N. C.	1849, Oct. 31.	18	3.63	Yale Coll.; C. U. Shepard, et
Petersburg, Tenn. .	1855, Aug. 5.	4	3.20	C. U. Shepard, et al. [al.
Harrison Co., Ind. .	1859, Mar. 28.	2	3.46	C. U. Shepard, et al.
Bethlehem, N. Y. .	1859, Aug. 11.	$\frac{1}{2}$	3.56	Albany Cabinet, et al. [al.
New Concord, O. .	1860, May 1.	700	3.54	Marietta Coll.; Yale Coll., et

EXPLANATION OF THE TABLES.

Table I., page 251, contains a comparison of French millimetres with English inches, and will be found convenient for reducing French measures into English. It is deduced from the assumption that the French metre at the freezing point is equal to 39.37079 English inches at the temperature of 62° Fahrenheit, the standard temperature of the French scale being 32° Fahrenheit; and that of the English scale being 62° Fahrenheit. This is the result given by Captain Kater in the Philosophical Transactions for 1818, page 109. The table of proportional parts in the last column gives the value of tenths of a millimetre in English inches, and will serve for hundredths by removing the decimal point one place to the left.

Table II., page 252, enables us to convert French metres into English feet, and is derived from the same data as the preceding table; that is, the French metre is equal to 3.2808992 English feet. The table of proportional parts in the last column may be used in the same manner as described in Table I.

Table III., page 253, enables us to convert French kilometres into English miles, and is derived from the same data as Table I.; that is, the French kilometre is equal to 0.6213824 English mile. The table of proportional parts in the last column may be used in the same manner as described in Table I.

Table IV., page 254, enables us to convert French feet into English feet. The old legal standard of France was the *Toise de Pérou*, so called from its being used by the French academicians in their measurement of an arc of the meridian in Peru. It is formed of iron, and was made in 1735. According to Base du Système Metrique, t. iii., p. 237, the metre is equal to 0.513074 toise, or 3.078444 French feet, which is equal to 3.2808992 English feet. Hence one French foot is equal to 1.065765 English feet.

The arrangement of Table IV. is similar to that of the preceding tables. The same table will serve equally well for converting French inches into English inches.

Table V., page 255, enables us to convert degrees of the centesimal thermometer into degrees of Fahrenheit. It is founded on the equation $x°$ centesimal$=(32°+\frac{9}{5}x°)$ Fahrenheit.

Table VI., page 256, enables us to convert degrees of Reaumur's thermometer into degrees of Fahrenheit. It is founded on the equation $x°$ Reaumur$=(32°+\frac{9}{4}x°)$ Fahrenheit.

Table VII., page 257, gives the height of a column of air corresponding to a tenth of an inch in the barometer for different temperatures from 40° to 90°, and may be used for reducing barometrical observations to the level of the sea, or to any other level.

Example. At Cambridge, Massachusetts, at 70 feet above the sea, the mean height of the barometer is 29.940 inches, and the mean temperature 48°; what would be the height at the level of the sea?

From Table VII. we find for barometer 29.94, and temperature 48°, the number 90.8.

Then the required correction equals $\frac{70}{908}=0.075$.

And 29.940+.075=30.015 inches, the height of the barometer at the level of the sea.

This table is derived from Guyot's Meteorological Tables, published by the Smithsonian Institution, D. 92.

Table VIII., pages 258-9, gives the correction to be applied to English barometers with brass scales for reducing the observations to 32° Fahrenheit, and is the same as adopted by the Royal Society of London. From 29° up the correction must be *subtracted* from the observed height, while from 28° down it must be added.

Example 1. Observed height of barometer, 29.876; attached thermometer, 73° Fahrenheit.

On page 259, in the column headed 30 inches, on the horizontal line corresponding with 73° in the first vertical column, we find the correction —.119. Hence the barometer, reduced to 32° Fahrenheit, will be 29.876—.119=29.757 inches.

Example 2. Observed height of barometer, 29.854; attached thermometer, 17° Fahrenheit.

On page 258, under 30 inches and opposite to 17°, we find the correction +.031. Hence the barometer, reduced to 32° Fahrenheit, will be 29.854+.031=29.885 inches.

If we wish the correction for a fraction of a degree, we must take a proportional part of the difference between the corrections for the nearest whole degrees in the table.

Table IX., pages 260–1, enables us to compute the difference in the heights of two places by means of the barometer. The construction of the table is fully explained in my Introduction to Practical Astronomy, page 480.

Method of Computation.

Take from Part I., page 260, the two numbers corresponding to the observed barometric heights h and h'. From their difference subtract the correction found in Part II., with the difference $T-T'$ of the thermometers attached to the barometers. We thus obtain an approximate altitude, a.

We then calculate the correction $\dfrac{t+t'-64}{900}a$ for the temperature of the air by multiplying the nine hundredth part of a by the sum of the temperatures t and t' diminished by 64. This correction is of the same sign as $t+t'-64$. We thus obtain a second approximate altitude, A.

With A and the latitude of the place, we seek in Part III. the correction arising from the variation of gravity with the latitude. With A we also seek in Part IV. the correction arising from the diminution of gravity on a vertical. Also, when the height of the lower station is considerable, another small correction is found in Part V. The last two corrections are always additive.

Example. The following observations were made at Geneva and on Mount Blanc, 3.3 feet below the summit of the mountain.

Mount Blanc, $h'=16.695$ inches, $T'=24°.4$ Fah., $t'=18°.3$ Fah.
Geneva, $h=28.727$ " $T=65.5$ " $t=66.7$ "

Part I. gives $\begin{cases} \text{for } h=28.727 \text{ inches,} & 26476.8 \\ \text{for } h'=16.695 \text{ "} & 12297.3 \end{cases}$

Difference, $\overline{14179.5}$

Part II. gives for $T-T'=41°.1$ -96.2

Approximate altitude, $a=\overline{14083.3}$

$t+t'-64=21.$

$$\frac{21a}{900} = +328.6$$

Second approximate altitude, $A=14411.9$
Part III. gives for lat. 46°, -1.4
Part IV. gives for 14412, $+46.0$
Part V. gives for bar. 28.7, $+1.5$

Sum, $\overline{14458.0}$

Height of Geneva above the sea,   1335.4
Barometer below summit of Mt. Blanc,   3.3
Height of Mount Blanc above the sea, $\overline{15796.7}$ feet.

Table X., page 262, furnishes the mean height of the barometer at nine stations upon the American continent; also at nine stations in the western part of the Eastern continent; and at nine stations in the eastern part of that continent. The precise locality of these stations is shown in the following table:

	Latitude.	Longitude.		Latitude.	Longitude.
Georgetown, Br. Guiana	6° 50'	58° 12'	Greenwich, England..	51° 28'	0° 0'
Havana, Cuba.............	23 9	82 23	St. Petersburg, Russia	59 56	- 30 18
Natchez, Miss.	31 34	91 24	Archangel, Russia.....	64 32	- 40 33
St. Louis, Mo.	38 37	90 15	Hammerfest, Lapland	70 39	- 23 42
Philadelphia, Penn.	39 58	75 10	Singapore, Malacca...	1 17	-103 50
Boston, Mass.	42 21	71 3	Madras, Hindostan ...	13 4	- 80 19
Toronto, Canada	43 40	79 22	Bombay, Hindostan ..	18 56	- 72 54
Port Bowen, Arc. Reg..	73 14	88 56	Canton, China	23 8	-113 16
Van Rensselaer Harbor.	78 37	70 53	Benares, Hindostan..	25 18	- 82 56
Christiansborg, Africa..	5 24	- 0 16	Pekin, China...........	39 54	-116 26
Aden, Arabia	12 50	-45 6	Tiflis, Georgia	41 41	- 45 17
Cairo, Egypt	30 2	-31 15	Nertschinsk, Russia...	51 18	-119 20
Constantinople, Turkey.	41 0	-29 0	Jakutsk, Siberia.......	62 1	-129 44
Paris, France	48 50	- 2 20			

A portion of the numbers in this table was derived from an article by Professor Dove, published in the Monatsberichte der Akademie zu Berlin, 1860, pages 644–692; the remainder was derived from a variety of sources.

Table XI., page 263, furnishes the mean height of the barometer for all hours of the day at nine stations from the equator to latitude 78°. Most of these places are included in the preceding list. A portion of these numbers was derived from Kämtz's Lehrbuch der Meteorologie, vol. ii., pages 254–259 ; the others were derived from various sources.

Table XII., page 263, furnishes the depression of mercury in glass tubes on account of capillarity, according to several different authorities.

Table XIII., page 264, gives the weight of a cubic foot of dry air and of saturated air under a barometric pressure of 30 inches, at temperatures between 0° and 90° F. The weight of a cubic foot of dry air is assumed to be 563 grains troy at a temperature of 32° F., and the coefficient of expansion is 0.002083 of its bulk for 1° F.

The weight of a cubic foot of saturated air is found by adding to the weight of a cubic foot of dry air the weight of a cubic foot of vapor, and correcting this result for the enlargement of volume resulting from the mixture. This table is derived from the Greenwich Meteorological Observations for 1842, pages 46 and 51.

Table XIV., page 265, shows the height of the barometer corresponding to temperatures of boiling water from 188° to 213° F. The temperature at which water boils in the open air depends upon the weight of the atmospheric column above it, and under a diminished barometric pressure the water will boil at a lower temperature. Since the weight of the atmosphere decreases with the elevation, it is evident that, in ascending a mountain, the higher the station, the lower the temperature at which water boils. Hence, if we know the height of the barometer corresponding to the temperature of boiling water, we can measure the altitude of a mountain by observing the temperature at which water boils. This table is copied from my Practical Astronomy, page 398.

Table XV., page 266, gives the corrections to be applied to the means of the hours of observation to obtain the true mean temperature at New Haven. These numbers are the differences, with opposite signs, between the hourly temperatures and the

T

true mean temperature of each month and also of the year. Thus, at New Haven, the mean temperature of January is 26°.5; the mean temperature at midnight in January is 24°.2; the difference is 2°.3, which is the quantity which must be added to midnight observations to obtain the mean temperature of that month, and so for the other hours and months of the table.

At the bottom of the table is given a comparison of some of the different modes which have been proposed for deducing the mean temperature from a limited number of observations. Thus, if we have observations at 7 A.M. and 1 P.M. in January, the former require a correction of +4°.4 and the latter of −6°.1; the mean of the two will require a correction of −0°.8, as given in line 26th of the table.

If we have observations at 6 A.M., 2 and 6 P.M. in January, the corrections for these three hours will be +4°.3, −6°.3, and −1°.4. The mean correction is −1°.1, which is the number given in line 36th of the table.

If we have observations at 7 A.M., 2 and 9 P.M, and if we add twice the nine o'clock observation to the sum of the other two observations, and divide the result by 4, the error of the result for the separate months in only one instance exceeds a quarter of a degree. This table is copied from the Transactions of the Connecticut Academy of Arts and Sciences, vol. i., p. 231.

Table XVI., page 267, is constructed for Greenwich, England, in the same manner as the preceding, and is taken from the Greenwich Meteorological Observations.

Table XVII., pages 268-9, gives the mean temperature of 45 places on the American continent for each month of the year. Some of these numbers are derived from Dove's Tables in the Report of the British Association for 1847, page 376; others are derived from the Army Meteorological Register, 1855, and some from other sources.

Table XVIII., page 270, furnishes a list of places whose mean temperature is above 80° F. The materials are derived chiefly from Dove's Tables.

Table XIX., page 270, furnishes a list of places whose mean

temperature is below 18° F. This is also derived chiefly, but not exclusively, from Dove's Tables.

Table XX., page 271, furnishes a list of places where the mean temperature of the hottest month differs less than six degrees from that of the coldest month, and is chiefly derived from Dove's Tables.

Table XXI., page 271, furnishes a list of places where the mean temperature of the hottest month differs more than sixty-six degrees from that of the coldest month. The materials are derived partly from Dove's Tables, partly from Kupffer's Annales, and partly from other sources.

Table XXII., page 272, furnishes a list of places where the annual range of temperature is less than 40°. The materials were derived partly from Arago's Works, vol. viii., pages 184–646, but many of the numbers were obtained by an extensive comparison of Meteorological Journals.

Table XXIII., page 272, furnishes a list of places where the annual range of temperature is greater than 130°. The materials were derived partly from Arago, vol. viii., but many of the numbers were obtained by an extensive comparison of Meteorological Journals, particularly Kupffer's Annales, the Army Meteorological Register, and the New York Meteorological Observations.

Table XXIV., page 273, shows the height of the line of perpetual snow above the level of the sea for a variety of latitudes. The works chiefly depended upon in preparing this table are the Encyclopædia Metropolitana, Art. Meteorology, page 84; Müller's Lehrbuch der Kosmischen Physik, page 353; and Kaemtz's Meteorology.

Table XXV., page 273, contains the factors by which the difference of readings of the dry-bulb and wet-bulb thermometers must be multiplied in order to produce the difference between the readings of the dry-bulb and dew-point thermometers. These factors are derived from a long series of observations made at the Greenwich Observatory, and enable us to convert observations made with the wet-bulb thermometer into observations made with Daniell's hygrometer.

Example. The temperature of the air being 44°.5, and that of the wet-bulb being 38°.7, it is required to determine the dew-point. The difference between the dry and wet bulb thermometer is 5°.8, which, multipled by 2.17, gives 12°.6, which is the difference between the dry-bulb and dew-point thermometers. Hence the dew-point was at 31°.9.

Table XXVI., pages 274–5, shows the relative humidity of the air at temperatures from 6° to 95°, and for a difference of temperature of air and of the dew-point from 0° to 24°. The relative humidity is the ratio of the quantity of vapor actually contained in the air to the quantity it could contain if fully saturated, Art. 105. This humidity is deduced from Table XXVII.

Example. Suppose the temperature of the air is 90° F., and that of the dew-point is 80° F., the difference being 10° F. According to Table XXVII., the elastic force of vapor at these two temperatures is 1.410 and 1.023; their ratio is .73, which is the relative humidity, and is the number given in the table for a temperature of 90°, and a dew-point 10° below the temperature of the air. Making the point of saturation 100, all the numbers in the table are to be regarded as integers. This table is abridged from one given in the Smithsonian Meteorological Tables, B. 75.

Table XXVII., page 276, gives the elastic force of aqueous vapor for temperatures from −30° to 101° F., according to the experiments of Regnault. The table is abridged from the Smithsonian Tables., B. 43.

Table XXVIII., page 277, is designed to furnish a comparison between the pressure· and velocity of the wind. It is derived from the Meteorological Papers of the British Board of Trade, third number, page 99, and was computed by Colonel James, assuming that the square of the velocity in miles per hour, multiplied by 0.005, gives the pressure in pounds per square foot. These numbers differ slightly from those given on page 70, but neither table can be regarded as perfectly reliable. More numerous experiments are needed for determining the pressure of the wind at different velocities.

Table XXIX., pages 278-9, gives the average amount of rain for each month·of the year at 45 stations on the American continent, extending from near the equator to the highest northern latitude for which such observations could be found. A considerable part of these results is taken from the Army Meteorological Register, published in 1855; the remainder are chiefly derived from Dove's Klimatologische Beiträge, and the Meteorological Observations of the Smithsonian Institution, while a few have been derived from other sources.

Table XXX., page 280, gives a list of stations at which the annual fall of rain is less than ten inches. The table furnishes the authorities for the results here given.

Table XXXI., page 280, gives a list of stations at which the annual fall of rain exceeds twelve feet. These stations have generally considerable elevation above the sea, but in many of the cases the heights, not being accurately known, could not be given in the table.

Table XXXII., page 281, shows the comparative radiating power of different substances at night, according to the observations of Mr. Glaisher, made at Greenwich, England, and published in the Philosophical Transactions for 1847, page 119. The numbers refer to long grass as the unit.

Table XXXIII., page 281, shows the fall of the barometer during several remarkable hurricanes in the West Indies, the East Indies, and elsewhere. The table shows the fall in the number of hours given in column fourth, but this is not generally the *entire* *fall* of the barometer during the day of the hurricane, for the·highest point of the barometer usually occurs some hours before the rapid fall begins, or some hours after the most rapid rise at the close of the storm.

Table XXXIV., page 282, gives a catalogue of auroras observed in Europe since 1685, and in America since 1742, the latter being chiefly confined to Boston and New Haven. These numbers show clearly the unequal frequency of auroras in the different years, and these inequalities indicate a period of ten or twelve years, with a more decided period of about sixty years. The table also shows the relative frequency of solar spots since 1749, and the mean daily range of the magnetic needle as observed in Europe since 1782. It will be noticed that the last two phenomena show most decided periodic inequalities, and these periods correspond remarkably with the periods of auroral abundance. The table is abridged from several tables published in the Smithsónian Report for 1865, pages 225–243.

Table XXXV., page 283, gives a catalogue of the principal iron meteors exceeding 40 pounds in weight. It is not claimed that this catalogue is complete, for in the report of many meteors the weight is not definitely stated. The number of iron meteors whose weight is less than 40 pounds is nearly equal to the number embraced in this catalogue. This catalogue has been compiled from a great variety of sources, but chiefly from Buchner's Meteoriten, 1863.

Table XXXVI., page 284, gives a list of the aerolites fallen in the United States. Besides these, there are five or six other cases in which aerolites have been claimed to have fallen, but as those cases are not considered to be sufficiently well attested they have been omitted.

EXPLANATION OF THE PLATES.

PLATE I. shows the prevalent winds at eight stations of the American continent from near the equator to latitude 78° N. Horizontal and vertical lines are drawn to represent the four cardinal points, and diagonal lines are drawn for the intermediate directions. On these eight lines are set off distances corresponding to the relative frequency of the winds from these eight quarters. The curve line passing through the eight points thus determined may be regarded as showing the prevalent wind at that station. It is thus seen that at Van Rensselaer Harbor and at Godthaab the prevalent wind is from the N.E.; at Norway House it is from the north; at St. Johns, New York, and Savannah the prevalent wind is from the S.W.; at Matanzas it is from the N.E.; and at Georgetown it is nearly from the east.

Plate II. represents the six varieties of cloud described on pages 101 and 102, each variety being indicated by a symbol shown at the bottom of the page.

Plate III. exhibits two outline maps of the United States, designed to illustrate the phenomena of a storm described on pages 142 and 143.

WORKS ON METEOROLOGY.

THE student who desires a more thorough knowledge of meteorology than can be obtained from this treatise, is referred to the following works and memoirs:

Annales de l'Observatoire Physique Central de Russie. One large quarto volume of observations annually.

Annuaire Magnetique et Meteorologique du Corps des Ingenieurs des Mines de Russie.

Annuaire de la Societe Meteorologique de France. One large octavo volume annually since 1849.

Apjohn. Theory of the Moist-bulb Hygrometer, Edinb. Philos. Transact., xvii.

Arago. Œuvres Completes, 12 volumes, 8vo. Etat thermometrique du Globe terrestre.—Influence de la Lune.—La pluie.—Le tonnere.—Puits Artesiens, etc.

Baddeley. On Dust Whirlwinds and Cyclones in India.

Biot. On Mirage and unusual Refraction, Mem. de l'Academie, 1809.

Birt. Reports to the British Assoc. on Atmospheric Waves, 1844–8.

Blodget. Climatology of the United States. Philadelphia, 1857.

Boué. Katalog der Nordlichter bis 1856. Wien Acad., 74 pages.

Bravais and Martin. Comparaisons Barom. faites dans le Nord de l'Europe, Mem. Acad. Brux., xiv.

Brewster. On the Mean Temperature of the Globe, Edinb. Phil. Transact., ix. Results of Thermometrical Observations at Leith Fort, do., x.

Buchan. Handy Book of Meteorology. London, 1867.

Buchner. Meteoriten. Leipsig, 1863.

Bulletin de l'Observatoire de Paris. One sheet daily, containing Meteorological Reports from every part of Europe.

Buist. Catalogue of Indian Hailstones and Meteors.

Buys Ballot. Sur la marche annuelle du thermometre et du barometre en divers lieux de l'Europe, 1849-59, Amsterd. Acad., 1861, 116 pages.

Cordier. On Temperature of Interior of the Earth, Mem. Acad. Sci., 1827.

Correspondence, Met. de l'Obs. Central Physique de Russie. A thin quarto volume annually.

Cotte. Meteorologie, Paris, 1774.

Daguin. Traité de Physique, avec les applications a la Meteorologie, 4 vols., Paris, 1861.

Dalton. On Rain and Dew, Manchester Mem., v. On the Constitution of Mixed Gases, do. Met. Obs. and Essays, London. On Constitution of the Atmosphere, Phil. Trans., 1826. On Height of Aurora Borealis, Phil. Trans., 1828.

Daniell. Meteorological Essays, 2 vols. 8vo, London. On the Constitution of the Atmosphere.

De Luc. On Hygrometry, Ph. Trans., 1791. On Evaporation, do., 1792.

Dove. Tables of Mean Temperature, Rep. of Br. Assoc., 1847. Verbreitung der Wärme auf der Oberfläche der Erde, 4to. Klimatologische Beiträge, 1861. Das Gesetz der Stürme, 2d edit.

Ermann. Ueber Boden und Quellen Temperatur. Ueber einige Barom. Beob., Poggendorff, lxxxviii.

Espy. Philosophy of Storms, Boston, 1841. Reports on Meteorology of U. S.

Fitzroy. Weatherbook, a Manual of Practical Meteorology. London, 1863.

Forbes. Report to Brit. Assoc. on Meteorology, 1832. Supplementary Report, 1840. On the Climate of Edinburg for 56 years, Trans. R. S. Edinb., xxii., part 2.

Fritsch. Periodische Erscheinungen in Wolkenhimmel, R. Bohem. Acad., v. Folge, Bd. iv.

Galton. Meteorographica, or Maps of the Weather, 4to. London, 1863.

Gehler. Wörterbuch, Arts. Meteorologie, Regen, etc.

Glaisher. On Nocturnal Radiation, Phil. Trans., 1847. On Correction of Monthly Means of Met. Obs., Phil. Trans., 1848.

Greg. On Aerolites, L., E., and Dub. Phil. Mag., 1854, p. 329. Catalogue of Meteorites, Rep. Br. Assoc., 1860, p. 48.

Guyot. Meteorological Tables, 8vo, 1859.

Harvey. Art Meteorology, Encyc. Metropolitana.

Heis. Ueber Sternschnuppen. Köln, 1849.

Herschel, J. F. W. Admiralty Manual of Scientific Inquiry. London, 1851. Meteorology, 1862.

Hopkins. On Winds and Storms. London, 1860.

Hough. New York Meteorological Observations, 1826–50. Albany, 1855.

Howard. Climate of London, 3 vols., 8vo. On a Met. Cycle of 18 years, Phil. Trans., 1841. Barometrographia.

Hudson. Hourly Obs. of the Barometer, Phil. Trans., 1832.

Humboldt. On Isotherms, Mem. d'Arceueil, iii. On Inferior Limit of Perpetual Snow, Ann. de Chim., xiv. Kosmos.

Jelinek. Tägliche Gang der Meteorolog. Elemente. Wien.

Johnson. Met. Obs. at Radcliffe Observatory, Oxford.

Johnston, Keith. Physical Atlas of Natural Phenomena.

Kämtz. Lehrbuch der Meteorologie. Leipsig, 3 vols. On Isobarometric Lines. Meteorology translated C. V. Walker.

Kirkwood. Meteoric Astronomy. Philadelphia, 1867.

Koller. Gang der Wärme in Oesterreich (Kremsmunster, 1841).

Kupffer. On Springs and Earth Temp., Poggendorff, xx.

Lamont: Beobachtungen auf d. Hohenpeissenberg, 1792–1850. München, 1851. Annalen fur Meteorologie.

Lawson. Army Meteor. Register for 12 years, 1843–54. Washington, 1855.

Loomis. On two Storms which occurred in February, 1842, Trans. Am. Phil. Soc., vol. ix. On the Storm of December, 1836, Smith. Contrib., 1860. On the Aurora Borealis, Smith. Report, 1865, p. 208. Mean Temperature of New Haven, Conn., Trans. Conn. Acad., vol. i.

Mahlmann. Temperature auf der Oberfläche der Erde (Dove's Repertorium, Bd. iv.).

Mairan. Traité de l'Aurore Boreale, 2d ed., Paris.

Maury. Storm and Rain Charts of the North and South Atlantic.

Meech. Relative Intensity of the Heat and Light of the Sun upon different Latitudes, Smith. Contr., 1855.

Meteorological Society. (of London). Transactions and Council Reports.

Mühry. Klimatologische Untersuchungen, 1858.

Müller. Lehrbuch du Kosmischen Physik.

Newton. On Shooting Stars, Mem. Nat. Acad. Sciences, vol. i. Original Accounts of November Star Showers, Am. Jour. Science, N. S., vol. xxxvii., p. 377. Contributions to Astro-Meteorology, Journ. Sc., vol. xliii., p. 285, etc.

Olmsted. Secular Period of the Aurora Borealis, Smith. Contr., 1855.

Partsch. Die Meteoriten. Wien, 1843...

Peltier. Sur les Trombes. Paris.

Phipson. Meteors, Aerolites, and Falling Stars. London, 1867.

Piddington. Nineteen Memoirs on Cyclones in the Indian and China Seas.—Sailor's Horn-Book.

Plantamour. Des Anomalies de la Temperature a Genève, 1867. Resumé des Obs. Therm. et Bar. a Genève.

Pouillet. Mem. sur la Chaleur Solaire, Comptes Rendus, 1838.

Quetelet. Sur le Climat de la Belgique. Catalogue des Apparitions des Etoiles Filantes, Acad. Brux., 1839. Variations Periodiques de la Tem., Acad. Brux., xxviii. Meteorologie de Belg.

Reid. Law of Storms. On Storms and Variable Winds.

Redfield. On the Courses of Whirlwinds. Am. Journ. Sc., xxxv., etc.

Robinson. Improved Anemometer, Royal Irish Academy, xxii.

Sabine. Report on Meteorology of Toronto, Br. Assoc., 1844. Lunar Tide at St. Helena, Phil. Trans., 1847. Meteorology of Bombay, Phil. Trans., 1853. Variations of Temperature at Toronto, Phil. Trans., 1853.

Saussure. Essais de l'Hygrometrie. 1783.

Schlagintweit. Results of a Scientific Mission to India, 1854–8, 4 vols. 4to, Leipsig.

Schouw. Beiträge zu Vergleichenden Klimatologie, Bibl. U., xxxiv.

Schubler. Atmospheric Electricity. Jahrbuch der Chem. und Phys., 1829.

Secchi. Results of Met. Obs. at Rome, Bibl. U., 1857.

Sykes. On Atmospheric Tides, Phil. Trans., 1835. Observations in India, Phil. Trans., 1850.

Thomson. Introduction to Meteorology. London 1849.

Welsh. Account of four Balloon Ascents, Phil. Trans., 1856.

Wells. On Dew. London 1818.

Whewell. On a new Anemometer, Trans. Cam. Phil. Soc., vi.

Wollaston. On the finite Extent of the Atmosphere, Phil. Trans.

Coffin. Winds of the Northern Hemisphere, Smithsonian Contributions, vol. vi.

Ferrel. Motions of Fluids and Solids relative to the Earth's Surface, Math. Monthly, vols. i. and ii.

Herrick. Register of the Aurora Borealis, Transactions of the Connecticut Academy, vol. i.

Quetelet. Meteorologie.

Smithsonian. Meteorological Results, 1854–9.

INDEX.

U

PLATE I.

VAN RENSSELAER HARBOR. *LAT. 78°37'*

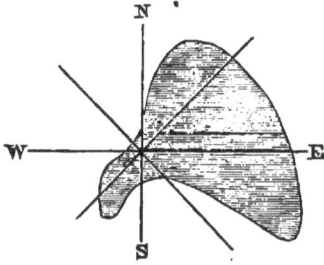

GODTHAAB, GREEN'D. *LAT. 64°10'*

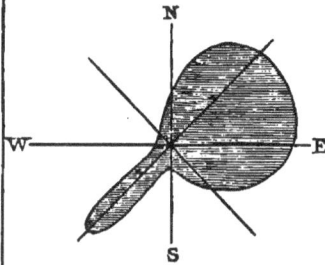

NORWAY HOUSE, H.B.TER. *LAT. 55°0'*

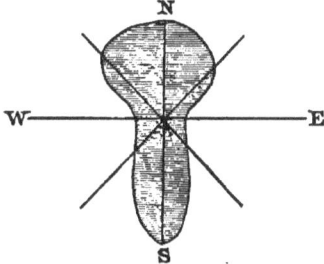

ST. JOHNS, NEWFOU'D. *LAT. 47°35'*

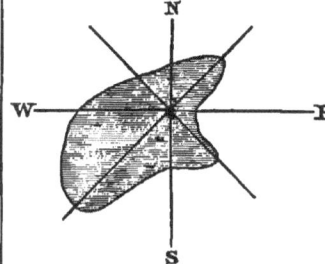

NEW YORK CITY. *LAT. 40°42'*

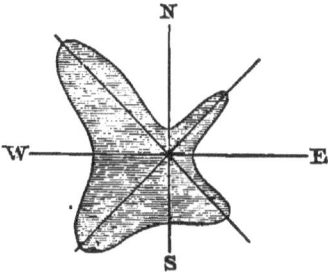

SAVANNAH, GA. *LAT. 32°6'*

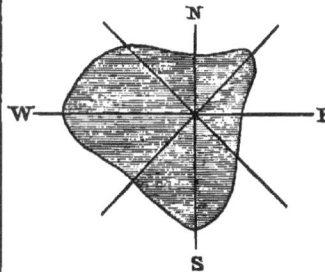

MATANZAS, CUBA. *LAT. 23°3'*

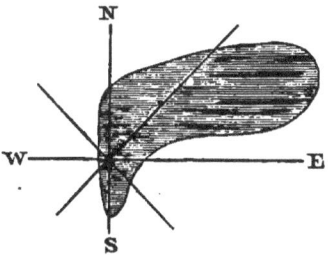

GEORGETOWN, B. G'A. *LAT. 6°49'N.*

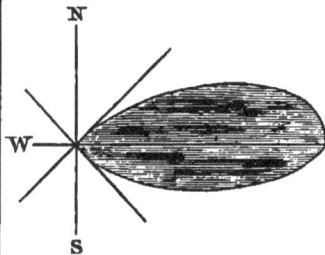

PLATE II.

Cirrus ・ Cirrocumulus ・ Cirrostratus
Cumulus ・ Cumulostratus ・ Stratus

PLATE III.

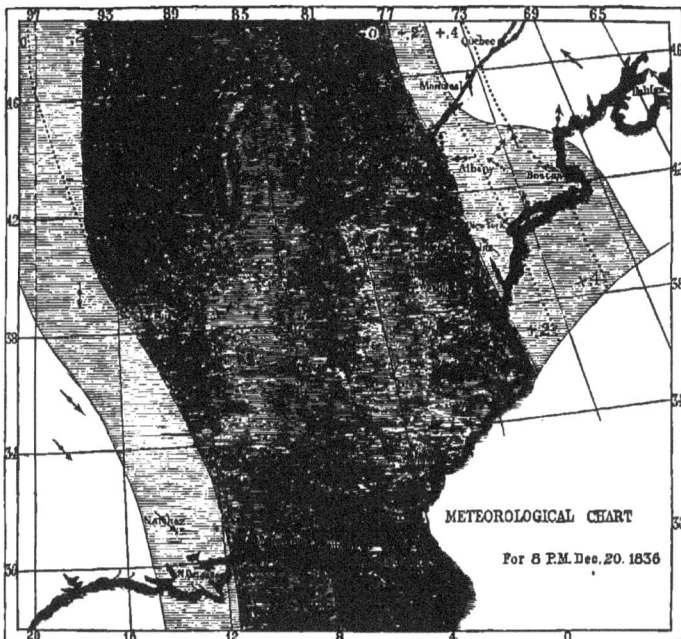

METEOROLOGICAL CHART

For 8 P.M. Dec. 20. 1836

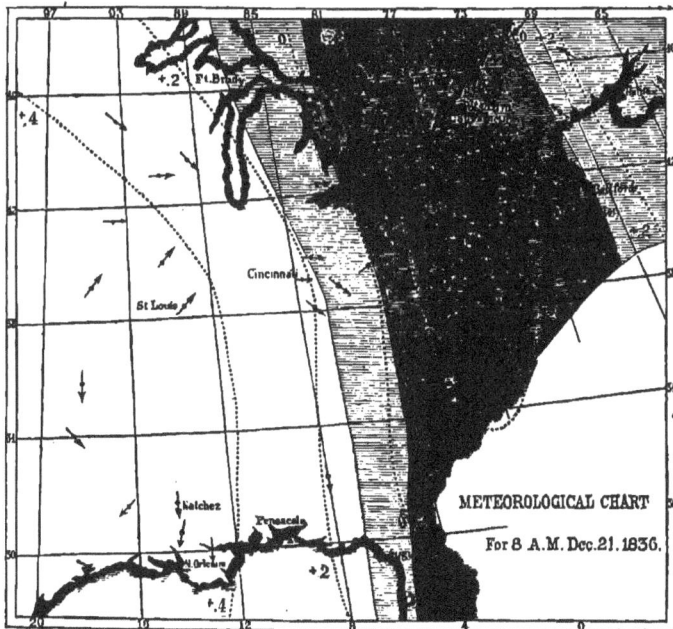

METEOROLOGICAL CHART

For 8 A.M. Dec. 21. 1836.